高等职业教育新形态系列教材

机电工程专业英语

主　编　朱允龙
副主编　谌　鑫　董丽娜

北京理工大学出版社
BEIJING INSTITUTE OF TECHNOLOGY PRESS

内容提要

本书根据高等职业院校机电一体化、机械设计和制造、数控技术、空调与制冷等专业的特点，结合不同区域机械行业的特点，采用教、学、练一体的模式编写。内容的编写紧密联系应用实际，并配有丰富的图片、习题以及参考译文。

本书共4个模块，包括机械设计、机械制造、数控技术、机电一体化技术、制冷与空调、民航机务、石油行业、建筑工程设备、医疗器械和其他设备，共10个单元。

本书可以作为高等职业院校、高等专业学校以及成人高等院校机械大类专业及相近专业的通用教材，也可作为其他相近专业学生和工程技术人员的参考用书。

版权专有 侵权必究

图书在版编目（CIP）数据

机电工程专业英语 / 朱允龙主编．－－北京：北京理工大学出版社，2023.12（2024.11 重印）
ISBN 978-7-5763-3217-9

Ⅰ．①机… Ⅱ．①朱… Ⅲ．①机电工程—英语—高等职业教育—教材 Ⅳ．① TH

中国国家版本馆 CIP 数据核字（2023）第 243760 号

责任编辑：高雪梅　　　　　　　　文案编辑：辛丽莉
责任校对：周瑞红　　　　　　　　责任印制：李志强

出版发行 / 北京理工大学出版社有限责任公司
社　　址 / 北京市丰台区四合庄路6号
邮　　编 / 100070
电　　话 /（010）68914026（教材售后服务热线）
　　　　　（010）63726648（课件资源服务热线）
网　　址 / http://www.bitpress.com.cn
版 印 次 / 2024年11月第1版第2次印刷
印　　刷 / 河北鑫彩博图印刷有限公司
开　　本 / 787 mm × 1092 mm　1/16
印　　张 / 18.5
字　　数 / 401千字
定　　价 / 56.50元

图书出现印装质量问题，请拨打售后服务热线，负责调换

前　言

为贯彻落实党的二十大精神，落实立德树人的根本任务，适应当前经济社会对智能制造行业高素质和技术技能人才的需求，深化产教融合、校企合作，推动人才培养模式改革及信息化教学革新，体现岗、课、赛、证综合的育人理念，旨在为智能制造行业培养更多掌握理论知识与技能、具备相应外语能力、德才兼备的高素质技术技能人才，更好地服务于区域智能制造产业，编写了本书。

本书作为职业院校机电大类专业的基础课教材，是对高职学生外语教学的夯实与专业教学的补充。本书编排符合高职学生"循序渐进"的学习习惯，即教学内容由浅入深、层层推进，最后达到综合掌握的效果。全书共4个模块，涵盖机械设计、机械制造、数控技术、机电一体化技术、制冷与空调、民航机务、石油行业、建筑工程设备、医疗器械和其他设备，共10个单元。本书不追求大篇幅和高难度，而是根据智能制造大类专业的特点，提取出必要的基础知识点，进行章节设定，章节内容的生词量适中、篇幅长短得当、难易程度契合学生的接受能力。

本书编写针对性、实用性强，体现机电工程学科的特点，对该学科的教学内容，尽量做到完全覆盖。书中各文章篇幅内容充分但又不显得冗长。本书正文部分添加了相关阅读材料，教师可以根据教学情况灵活选读，每一篇都附有参考译文、长难句分析、单词释义及练习题，便于学生及时理解。

为贯彻落实《国务院办公厅关于深化高等学校创新创业教育改革的实施意见》（国办发〔2015〕36号）和《国务院办公厅关于深化产教融合的若干意见》（国办发〔2017〕95号）精神，深化产教融合、校企合作，本书编撰之初，编者结合区域用人需求，积极联系区域企业，沟通该教材开发意见及建议，设定了模块3典型机电行业、模块4典型机电设备的选学内容，让学生在学习模块1机械设计和制造、模块2电气控制技术，打好专业英语基础的同时，还能提升自己对专业英语的整体认知，并感受企业应用型英语的难度，培养在不同场景下专业英语的认知、应用机电工程专业英语的能力。

本书由朱允龙担任主编，由谌鑫、董丽娜担任副主编。本书具体编写分工如下：朱允龙主要编写模块2、模块4及课程思政部分；谌鑫主要编写模块1；董丽娜主要编写模块3。

编写本书时曾参阅了相关文献资料，在此，谨向文献的作者们深表谢意；同时还要感谢山东青州市南张石油机械厂、铁福来装备制造集团股份有限公司、中航工业贵航股份永红散热器公司、光达传动股份有限公司在本书编撰过程中给予的大力支持。

由于编者学识水平有限，书中难免存在错漏之处，恳请各位读者批评指正。

编　者

Contents

Module 1 Mechanical Design and Manufacturing ············ 001

   Unit 1   Mechanical Design ············ 002
      1.1   Introduction to Mechanical Design ············ 002
      1.2   Engineering Drawing ············ 007
      1.3   Computer Aided Design ············ 017
      1.4   Mould Design and Manufacturing ············ 025
   Unit 2   Machine Manufacturing ············ 033
      2.1   Introduction to Material Forming ············ 033
      2.2   Metal Cutting Processes ············ 043
      2.3   Milling Operations ············ 048
      2.4   Heat Treatment of Metal ············ 057

Module 2 Electrical Control Technology ············ 067

   Unit 1   Numerical Control Technology ············ 068
      1.1   Application of NC Technology ············ 068
      1.2   Numerical Control Machines ············ 074
      1.3   NC Programming ············ 092
   Unit 2   Electromechanical Integration Technology ············ 107
      2.1   Introduction to Mechatronics ············ 107
      2.2   Single Chip Microcomputer ············ 122
      2.3   Programmable Logic Controller ············ 126
      2.4   Sensor Technology ············ 131
      2.5   CAD/CAM ············ 137
      2.6   CAE/CIM ············ 145

Module 3 Typical Electromechanical Industry — 155

Unit 1 Refrigeration & Air-Conditioning — 156
1.1 Thermal Storage — 156
1.2 Tools and Equipment — 161
1.3 Heat Exchanger — 166

Unit 2 Civil Aviation Maintenance — 173
2.1 Aerodynamics Basics — 173
2.2 Airframe Construction — 182
2.3 Basic Knowledge of Ground Handling and Servicing — 189

Unit 3 Oil Industry — 203
3.1 Fundamentals of Quantitative Log Interpretation — 203
3.2 Drilling Platform and Drill Pipe Lifting Equipment — 216

Module 4 Typical Electromechanical Equipment — 227

Unit 1 Construction Equipment — 228
1.1 Slush Supply Pump — 228
1.2 Centrifuge — 235

Unit 2 Medical Apparatus and Instruments — 244
2.1 Monitor — 244
2.2 Defibrillator — 260

Unit 3 Other Equipment — 271
3.1 Tunnel Construction Equipment — 271
3.2 Vehicle Radiator — 279

References — 287

目 录

模块 1 机械设计和制造 ... 001

 单元 1 机械设计 ... 002
  1.1 机械设计概论 ... 002
  1.2 工程制图 ... 007
  1.3 计算机辅助设计 ... 017
  1.4 模具设计和制造 ... 025

 单元 2 机械制造 ... 033
  2.1 材料成型概论 ... 033
  2.2 金属切削工艺 ... 043
  2.3 铣削加工 ... 048
  2.4 金属热处理 ... 057

模块 2 电气控制技术 ... 067

 单元 1 数控技术 ... 068
  1.1 数控技术应用 ... 068
  1.2 数控机床 ... 074
  1.3 数控编程 ... 092

 单元 2 机电一体化技术 ... 107
  2.1 机电一体化概论 ... 107
  2.2 单片机 ... 122
  2.3 可编程逻辑控制器 ... 126
  2.4 传感器技术 ... 131
  2.5 CAD/CAM ... 137
  2.6 CAE/CIM ... 145

模块 3　典型机电行业 ... 155

单元 1　制冷与空调 ... 156
1.1　热储存 ... 156
1.2　工具与设备 ... 161
1.3　换热器 ... 166

单元 2　民航机务 ... 173
2.1　空气动力学基础 ... 173
2.2　机体构造 ... 182
2.3　地勤服务基本知识 ... 189

单元 3　石油行业 ... 203
3.1　定量测井解释基础 ... 203
3.2　钻井平台及其升降装置 ... 216

模块 4　典型机电设备 ... 227

单元 1　建筑工程设备 ... 228
1.1　供浆泵 ... 228
1.2　离心机 ... 235

单元 2　医疗器械 ... 244
2.1　监护仪 ... 244
2.2　除颤仪 ... 260

单元 3　其他设备 ... 271
3.1　隧道施工设备 ... 271
3.2　车载散热器 ... 279

参考文献 ... 287

Module 1

Mechanical Design and Manufacturing

Focus

- Unit 1   Mechanical Design
- Unit 2   Machine Manufacturing

Unit 1 Mechanical Design

Unit objectives

【Knowledge goals】
1. 掌握机械部件、机械设计流程等概念。
2. 掌握坐标系、视图类型、配合等概念。
3. 掌握 CAD 制图的工作内容、交换标准、CAD 系统的组成等。
4. 掌握压缩成型、注塑成型、铸造成型等概念。

【Skill goals】
1. 能够阅读专业英语文章。
2. 能够翻译专用英语词汇。

【Quality goals】
1. 在学习过程中发扬科学研究精神。
2. 增强团队合作意识。

1.1 Introduction to Mechanical Design

Text

机械设计

Mechanical design is either to formulate an engineering plan for the satisfaction of a specified need or to solve an engineering problem. It involves a range of disciplines in materials, mechanics, heat, flow, control, electronics and production.

Machinery Components

The major part of a machine is the mechanical system, and the mechanical system is decomposed into mechanisms, which can be further decomposed into mechanical components. In this sense, the mechanical components are the fundamental elements of machinery. On the whole, mechanical components can be classified as universal and special components. Bolts, gear, and chains are the typical examples of the universal components, which can be used extensively in different machines across various industrial sectors.

Mechanical Design Process

Product design requires much research and development. Many concepts of an idea must be studied, tried, refined, and then either used or discarded. Although the contents of each engineering problem is unique, the designers follow the similar process to solve the problems. The complete design process is often outlined in Fig. 1-1.

Recognition of Need Sometimes, design begins when a designer recognizes a need and decides to do something about it. The need is often not evident at all;

recognition is usually triggered by a particular adverse circumstance or a set of random circumstances, which arise almost simultaneously. Recognition of need usually consists of an undefined and vague problem statement.

Definition of Problem Definition of problem is necessary to fully define and understand the problem, after which it is possible to restate the goal in a more reasonable and realistic way than the original problem statement. Definition of the problem must include all the specifications for the thing that is to be designed. Obvious items in the specifications are the speeds, feeds, temperature limitations, maximum range, expected variation in the variables, and dimensional and mass limitations.

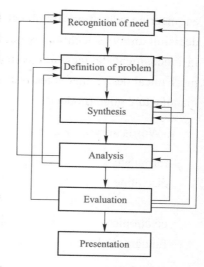

Fig. 1-1　Design process

Synthesis The synthesis is one in which as many alternative possible design approaches are sought, usually without regard for their value or quality. This is also sometimes called the ideation and invention step in which the largest possible number of creative solutions is generated. The synthesis activity includes the specification of material, addition of geometric features, and inclusion of greater dimensional detail to the aggregate design.

Analysis Analysis is a method of determining or describing the nature of something by separating it into parts. In the process, the elements or nature of the design are analyzed to determine the fit between the proposed design and the original design goals.

Evaluation Evaluation is the final proof of a successful design and usually involves the testing of a prototype in the laboratory. Here we wish to discover if the design really satisfies the needs.

The above description may give an erroneous impression that this process can be accomplished in a linear fashion as listed. On the contrary, iteration is required within the entire process, moving from any step back to any previous step.

Presentation Communicating the design to others is the finial, vital presentation step in the design process. Basically, there are only three means of communication. These are the written, the oral, and the graphical forms. A successful engineer will be technically competent and versatile in all three forms of communication. The competent engineer should not be afraid of the possibility of not succeeding in a presentation. In fact, the greatest gains are obtained by those willing to risk defeat.

Contents of Mechanical Design

Mechanical design is an important technological basic course in mechanical engineering education. Its objective is to provide the concepts, procedures, data, and decision analysis techniques necessary to design machine elements commonly found in mechanical devices and systems; to develop engineering students' competence of mechanical design that is the primary concern of machinery manufacturing and the key to manufacture good products.

【 Words and phrases 】

formulate	v. 制订；规划；确切表达；认真阐述；用公式表示
discipline	n. 纪律；惩罚；训导；科目；学科
	v. 惩罚；处分
electronics	n. 电子学；电子设备；电子器件
fundamental	adj. 根本的；基本的；必需的
	n. 基本原理；基音；基频
bolt	n. [机]螺栓；[机]螺钉；膨胀锚钉；角钢螺丝
gear	n. 排挡；齿轮；装备；服装
	v.（使）变速；（使）调挡
chain	n. 链；链条；一连串；一系列
	v. 用锁链拴住
universal component	通用组件
concept	n. 概念；观念
	adj.（围绕）某主题的
discard	v. 扔掉；弃置
	n. 被抛弃物
trigger	n. 扳机；起因；诱因
	v. 引发；激发
simultaneously	adv. 同时地
identification	n. 辨认；识别；确认
undefined	adj. 不明确的；未下定义的
vague	adj. 不明确的；不清楚的；（形状）模糊的
restate	vt. 重申；重新叙述；重讲
synthesis	n. 综合；综合体；（化学物质的）合成；音响合成
invention	n. 发明物；发明；创造
dimensional	adj. 维度的
aggregate	n. 总数；合计
	adj. 总计的，合计的；v. 集合，聚集
prototype	n. 原型；典型
	v. 制作（产品的）原型
erroneous impression	错误的印象
graphical form	图标形式

Course practice

1. Complex sentence analysis.

(1) Mechanical design is either to formulate an engineering plan for the satisfaction of a specified need or to solve an engineering problem.

① either...or...：或……或……。
② formulate：明确地表达，阐明。

【译文】机械设计用于阐述满足某种特殊需要的工程计划或解决具体的工程问题。

(2) ...the mechanical system is decomposed into mechanisms, which can be further decomposed into mechanical components.

① be decomposed into...：被分解为……
② which 引导一个定语从句，在从句中做主语，指前面的 mechanisms。

【译文】……机械系统可以分解为机构，机构又可以进一步分解为机械部件。

2. Translate the following paragraph.

The practice of design can be one of the most exciting and fulfilling activities that an engineer can undertake. There is a strong sense of satisfaction and pride in seeing the results of one's creative efforts emerge into actual products and processes that benefit people. To do design well requires a number of characteristics.

3. Choose the proper answer to fill in the blank and translate the sentences.

(into, of, at, as, of, within)

(1) The major part (　　) a machine is the mechanical system.

(2) The mechanical system is decomposed (　　) mechanisms, which can be further decomposed into mechanical components.

(3) Mechanical components can be classified (　　) universal and special components.

(4) The need is often not evident (　　) all.

(5) On the contrary, iteration is required (　　) the entire process, moving from any step back to any previous step.

(6) The competent engineer should not be afraid (　　) the possibility of not succeeding in a presentation.

Translation of the text

机械设计概论

机械设计用于阐述满足某种特殊需要的工程计划或解决具体的工程问题，它涉及材料、力学、热力、流体、控制、电子和生产等一系列学科。

机械部件

机器的主要组成部分是机械系统，机械系统可以分解为机构，机构又可以进一步分解为机械部件。从这个意义上说，机械部件是机械的基本要素。总体来说，机械部件可以分为通用部件和专用部件。螺栓、齿轮和链条是通用部件的典型例子，可广泛用于各种工业部门的不同机器中。

机械设计流程

产品设计需要大量的研究和开发。一个想法的许多概念必须经过研究、尝试、提炼，然后要么被使用，要么被抛弃。虽然每个工程问题的内容都是独特的，但设计师遵循相似的流程来解决问题。完整的流程通常如图 1-1 所示。

识别需求 有时，当设计师辨别出需求并决定为此做些什么时，设计就开始了。这种需求往往根本不明显；认知通常是由一个特定的不利环境或一组随机环境触发的，这些环境几乎同时出现。需求的识别通常由一个未定义且模糊的问题陈述组成。

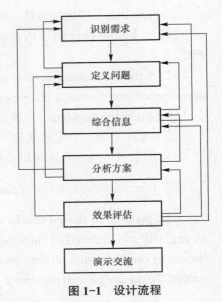

图 1-1 设计流程

定义问题 定义问题对于充分定义和理解问题是必要的，在此之后，可以比原始问题陈述更合理、更现实的方式重申目标。问题的定义必须包括所设计物的所有规格。规格中明显的条目是速度、进料、温度限制、最大范围、变量的预期变化以及尺寸和质量限制。

综合信息 综合信息是指寻求尽可能多的可替代设计方法，通常不考虑它们的价值或质量。这有时也被称为构思和发明阶段，其中产生了尽可能多的创造性解决方案。合成行为包括材料的规格、几何特征的添加，以及在总体设计中包含更多的尺寸细节。

分析方案 分析方案是一种通过将事物分成各个部分来确定或描述事物本质的方法。在这个过程中，对设计的元素或性质进行分析，以确定提出的设计与原始设计目标之间的契合度。

效果评估 效果评估是成功设计的最终证明，通常包括在实验室对原型进行测试。在这个过程中，我们希望发现设计是否真的满足需求。

上述描述可能会给人一种错误的印象，即该过程可以按所列的线性制造完成。相反，整个过程中需要迭代，可以从任何步骤返回到之前任意一步。

演示交流 与他人交流设计是设计流程中的最后一步，演示交流是最重要的一步。基本上，只有三种交流方式，即书面的、口头的和图形的形式。一个成功的工程师在技术上是有能力的，在这三种形式的交流中都是可以驾驭的。有能力的工程师不应害怕在演示中出现失败的可能性。事实上，那些愿意冒失败风险的人更容易得到最大的收获。

机械设计内容

机械设计是机械工程教育中一门重要的技术基础课。其目标是设计并提供机械设备和系统中常见的机器元素所需要的概念、程序、数据和决策分析技术；培养工科学生的机械设计能力，这是机械制造的首要问题，也是制造好产品的关键。

Evaluate

任务名称		机械设计概论	姓名	组别	班级	学号	日期	
考核内容及评分标准			分值	自评	组评	师评	平均分	
三维目标	知识	掌握机械部件、机械设计流程等概念	25分					
	技能	能够阅读专业英语文章、翻译专用英语词汇	40分					
	素养	在学习过程中发扬科学研究精神、增强团队合作意识	35分					
加分项	收获（10分）	收获（借鉴、教训、改进等）：	你进步了吗？			加分		
			你帮助他人进步了吗？					
	问题（10分）	发现问题、分析问题、解决方法、创新之处等：				加分		
总结与反思						总分		

1.2 Engineering Drawing

机械制造

Text

Engineering drawing is a professional basic subject, based on the projection theory of pictorial geometry, with ruler, compass and drawing board as tools, and blackboard, wood model and wall chart as media, with a history of more than 200 years. "Mechanical Engineering Drawing" is an introductory course that reflects the characteristics of engineering, and it is also one of the basic courses that engineering students must learn. It plays an important role in cultivating students' spatial imagination and ideation ability as the basis of creative thinking and promoting the industrialization process.

Coordinate System

In engineering drawing, the coordinate system is very important. The basic of all AutoCAD input is the Cartesian coordinate system, and the various input (absolute or relative) rely on this system. In addition, AutoCAD has two internal coordinate systems to help users keep track of where they are in a drawing: the world coordinate system (WCS) and the user coordinate system (UCS).

The fixed Cartesian coordinate system locates all points on an AutoCAD drawing by defining a series of positive and negative axes to locate positions in space. Fig. 1-2 (a) illustrates the axis for two dimension (2D) drafting. There is a permanent origin point (0, 0) which is referenced, an x axis running horizontally in a positive and negative direction from the origin, and a y axis traveling perpendicularly in a vertical direction. When a point is located, it is based on the origin point unless you are working in three dimensions, in which case you will have a third axis, called the z axis [Fig. 1-2 (b)].

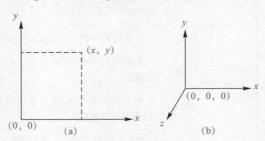

Fig. 1-2 The coordinate system
(a) Two dimensions; (b) Three dimensions

Types of Views

Many types of views are used to express the design ideas in engineering design in the area of engineering drawing.

Projection View A projection view is an orthographic projection of an object as seen from the front, top, right side, etc, as illustrated in the Fig. 1-3. Fig. 1-3 (a) is the Physical drawing of an object, and Fig. 1-3 (b) is the orthographic projection of the object.

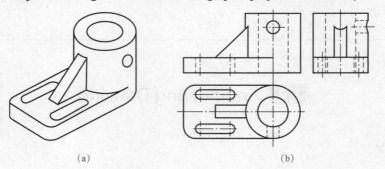

Fig. 1-3 An orthographic projection of an object
(a) Physical drawing of an object; (b) Orthographic projection of the object

Auxiliary View It is any view created by projecting 90° to an inclined surface, datum plane, or along an axis.

General View It is any view which is oriented by the user and is not dependent on any other view for its orientation.

Detailed View　It is any view which is derived by taking a portion of an existing view and scaling it for the purpose of dimensioning and clarification. Fig. 1-4 is a typical example of detailed view where there are two detailed views, Ⅰ and Ⅱ.

Revolved View　It is a view in which a cross-section is revolved 90° around the cutting plane line and offset alone, within a planar.

Full View　It is a view which shows the entire model.

Half View　It is a view which shows only the portion of the model on one side of a datum plane.

Broken View　It is used on large objects to remove a section between two points and move the remaining section close together.

Sectional View　It is a view which displays a cross-section for a particular view, as shown in Fig. 1-5.

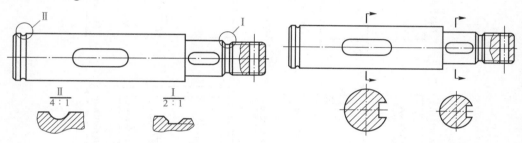

Fig. 1-4　Detailed view　　　　　　Fig. 1-5　Sectional view

Exploded View　The exploded view is a type of pictorial drawing designed to show several parts in their proper location prior to assembly. Although the exploded view is not used as the working drawing for the machinist, it has an important place in mechanical technology. Exploded views appear extensively in manuals and handbooks that are used for repair and assembly of machines and other mechanisms.

Partial View　When a symmetrical object is drafted, two views are sufficient to represent it (typically, one view is omitted). A partial view can be used to substitute one of the two views. Sectional and auxiliary views are also commonly used to present part detail. Sectional views are extremely useful in displaying the detailed design of a complicated internal configuration. If the section is symmetrical around a centerline, only the upper half needs to be shown, and the lower half is typically shown only in outline. Casting designer often employ sectional views to explode detail. When a major surface is inclined to three projection planes, only a distorted figure can be seen. An auxiliary plane that is parallel to the major surface can be used to display an undistorted view.

Mate

The mate between two mating parts is the relationship which results from the clearance or interference obtained. There are three classes of mate, namely, clearance, transition and interference. These conditions are shown in Fig. 1-6.

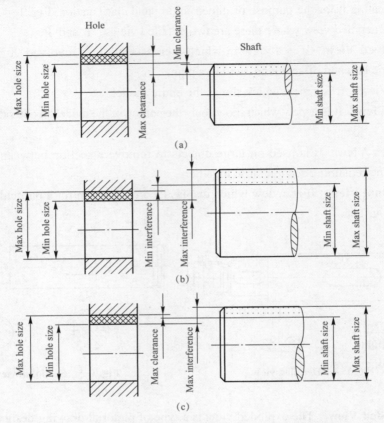

Fig. 1-6 Classes of mate between a hole and a shaft
(a) Clearance mate (the size of the shaft is always smaller than the hole);
(b) Interference mate (the size of the shaft is always larger than the hole);
(c) Transition mate (the limits are such that the condition may be of clearance of interference mate)

The Fig. 1-7 is a typical example of the dimensioning of mate code in the assembly drawing.

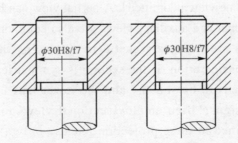

Fig. 1-7 A typical example of the dimensioning of mate code in the assembly drawing

1. Interchangeability

An interchangeable part is one which can be substituted for a similar part manufactured to the same drawing. The interchangeability of component parts is based upon these two functions:

(1) It is necessary for the relevant mating parts to be designed incorporating limits of size.

(2) The parts must be manufactured within the specified limits.

2. Limits of Size

In deciding the limits necessary for a particular dimension, there are three considerations: functional importance, interchangeability and economics.

It is necessary to have a knowledge of the purpose of this component, its replacement in the event of failure and avoidance of unnecessary time and money on production. The decision regarding to the degree of tolerance that can be utilized calls for discretion in the compromise between accuracy and economy. In order to assist the designer in his choice of limits and mates and to encourage uniformity throughout industry (home and abroad), a number of limit-and-mate systems have been published.

【Words and phrases】

coordinate	n. 坐标；坐标系
coordinate system	坐标系
Cartesian coordinate system	笛卡儿坐标系
axis	n. 坐标轴（pl. axes）
keep track of	跟踪于；定位于
world coordinate system (WCS)	世界坐标系
user coordinate system (UCS)	用户坐标系
perpendicularly	adv. 垂直地；直立地
projection	n. 估算；预测；投影；计划；设计
orthographic projection	正交投影
inclined surface	斜面
datum plane	基准面
general view	总图
orient	v. 朝向；面对；使适合；定向放置（某物）；确定方位
detailed view	n. 局部放大图
offset	n. 偏置；偏移
	v. 补偿；抵消
planar	adj. 平面的；二维的
sectional view	剖视图；剖面图
exploded view	分解图；爆炸图

partial	*adj.* 局部的；图示的
partial view	局部视图
symmetrical	*adj.* 对称的
mate	*v.* （使）交配；连接；配备
	n. <英，非正式>朋友；伙伴；同伴；同事
clearance	*n.* 审核批准；（飞机起降的）许可；准许；间距，间隙；清理
transition	*n.* 过渡；转变；（分子生物）转换
	v. 转变；过渡
interference	*n.* 干涉；干预；（收音机或电视机受到的）干扰信号；过盈

Course practice

1. Complex sentence analysis.

（1）The basic of all AutoCAD input is the Cartesian coordinate system, and the various input (absolute or relative) rely on this system.

① The basic of…：……的基础。

② rely on：依赖，依靠。

【译文】AutoCAD 全部输入均基于笛卡儿坐标系，各种输入方法（如绝对坐标、相对坐标）都依赖这个系统。

（2）When a point is located, it is based on the origin point unless you are working in three dimensions, in which case you will have a third axis, called the *z* axis...

① unless：引导让步状语从句。

② in which case：引导非限定性定语从句，在此作定语，修饰 dimensions。

③ called the *z* axis：作 third axis 的后置定语。

【译文】可根据原点（0,0）标记其他任意一个点，除非在三维空间中绘图时，应该具有称为 *z* 轴的第三个轴……

（3）Sectional and auxiliary views are also commonly used to present part detail.

to present part detail：作目的状语。

【译文】剖面图和辅助视图通常用于表达零件的局部细节。

（4）An interchangeable part is one which can be substituted for a similar part manufactured to the same drawing.

① interchangeable：作为动词 interchange 的形容词形式，译为可以互换的、可交换的。

② substituted for：代替，替换，取代。

【译文】可互换的零件是一种能由同一图纸加工出来的相似零件所代替的零件。

2. Answer the following questions according to the text.

(1) What is the Cartesian coordinate system and its function?

(2) How many types of views are there in our text? Please give their names.

(3) What is the detailed view and its function?

(4) How many types of mates are there in our text? Please give their names.

(5) What need to consider while deciding the limits necessary for a particular dimension?

3. Fill in the blanks with proper words or phrases according to the text (note the proper tense).

(1) AutoCAD has two internal coordinate systems, they are _____ and _____.

(2) There are many types of views in the text, namely _____, _____, _____, _____, _____, _____, _____, _____, _____, and _____.

(3) There are three classes of mates, namely _____, _____, and _____.

4. Translate the following phrases into Chinese according to the text.

(1) coordinate system
(2) cartesian coordinate axes
(3) keep track of
(4) world coordinate system (WCS)
(5) user coordinate system (UCS)
(6) orthographic projection
(7) projection plane
(8) partial view
(9) degree of tolerance
(10) home and abroad

5. Translate the following sentences.

(1) AutoCAD has two internal coordinate systems to help users keep track of where they are in a drawing: the world coordinate system (WCS) and the user coordinate system (UCS).

(2) The fixed Cartesian coordinate system locates all points on an AutoCAD drawing by defining a series of positive and negative axes to locate positions in space.

(3) There is a permanent origin point (0, 0) which is referenced, an x axis running horizontally in a positive and negative direction from the origin, and a y axis traveling perpendicularly in a vertical direction.

（4）It is necessary to have a knowledge of the purpose of this component, its replacement in the event of failure and avoidance of unnecessary time and money on production.

6. Write a 100-word summary according to the text.

Translation of the text

工程制图

工程制图是一门专业基础学科，以几何画法的投影理论为基础，以直尺、圆规、图板为工具，以黑板、木模、挂图为媒介，已有 200 多年的历史。"机械工程制图"是体现工科特点的入门课程，也是工科学生必须学习的专业基础课程之一。作为创造性思维的基础，它在培养学生的空间想象力和构思能力，以及促进工业化进程等诸多方面发挥重要的作用。

坐标系

在工程制图中，坐标系是十分重要的。AutoCAD 全部输入均基于笛卡儿坐标系，各种输入方法（如绝对坐标、相对坐标）都依赖这个系统。另外，AutoCAD 有两个内部坐标系：世界坐标系（WCS）和用户坐标系（UCS），用来帮助使用者确定所绘图形在绘图区中的位置。

固定的笛卡儿坐标系可以通过定义一系列用于确定空间位置的正负轴来标记 AutoCAD 图上的所有点。图 1-2（a）所示为用于二维绘图的坐标系，有一个作为参考点的固定原点（0，0），x 轴从原点出发沿着水平方向向正负方向延伸，y 轴从原点出发沿着垂直方向延伸。可根据原点（0，0）标记其他任意一个点，除非在三维空间中绘图时，应该具有称为 z 轴的第三个轴，如图 1-2（b）所示。

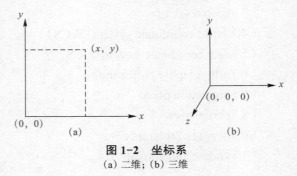

图 1-2　坐标系
(a) 二维；(b) 三维

视图类型

在工程制图领域，采用各种类型的视图来表达机械设计的设计理念。

投影视图　投影视图是从前面、顶面、右侧面等观察物体的正交投影图，如图 1-3 所示。图 1-3（a）为物体的实物，图 1-3（b）为物体的正交投影。

辅助视图　向倾斜面、参考面或沿着一个轴线作 90°投影所产生的视图即辅助视图。

总图　总图是由用户确定位置，并且其定向不依赖其他视图定位的视图。

局部放大图　为了标注尺寸和看清图形，从已知视图中取出一部分并将其放大的一种视图，即局部放大图。图 1-4 是一幅局部放大图的典型例子，图中有两处用到了局部放大图，分别为Ⅰ处和Ⅱ处。

旋转视图　平面中，横截面绕剖切线旋转90°后移出一定距离的视图为旋转视图。

全视图　全视图即显示整个模型的视图。

半视图　半视图只显示在参考面一侧的部分图形。

折断视图　折断视图用于尺寸大的物体，即移去（中间）两点间的一段截面并把留下的部分聚集到一起的截面视图。

剖面图　剖面图用于显示一个特定视图的横截面，如图1-5所示。

分解图　分解图是在装配前显示每个零件位置关系的一种示意图。尽管机械师不把分解图用作工作图，但它在机械技术上具有重要位置。分解图广泛出现在机器及其他机构维修和装配的说明书和手册中。

局部视图　当画一个具有对称结构的物体时，两个视图便足以表达（通常一个视图被省略），局部视图可用于代替视图的其中之一。剖面图和辅助视图通常用于表达零件的局部细节。剖面图用于显示内部结构复杂的细节，设计时尤其有用。如果截面沿着中心线对称，只展示上半部，下半部通常只用轮廓线显示出来。铸件设计师通常利用剖面图来展示细节。当一个主要面倾斜出三个投影面时，只能看到歪曲的图形。一个平行于该主要面的辅助平面，可用来显示物体未被歪曲的视图。

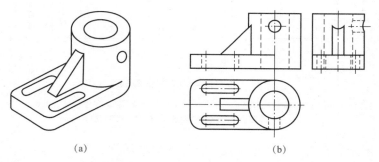

图1-3　物体的正交投影图

(a) 物体的实物图；(b) 物体的正交投影

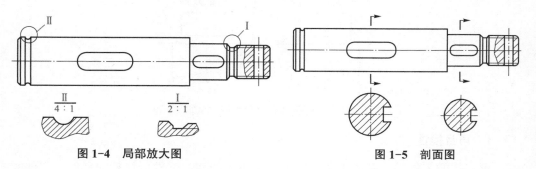

图1-4　局部放大图　　　　　图1-5　剖面图

配合

两个相互匹配零件之间的配合，是一种有间隙或过盈的关系。配合有三种等级，即间隙配合、过盈配合及过渡配合。这些条件如图1-6所示。

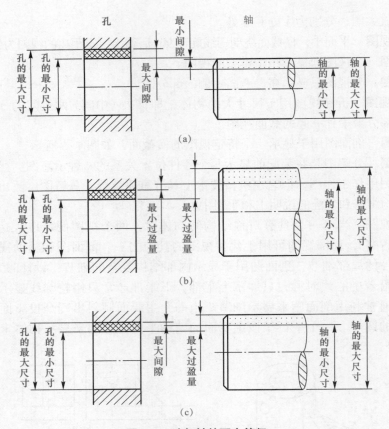

图 1-6 孔与轴的配合等级

(a) 间隙配合（轴的尺寸总是小于孔）；(b) 过盈配合（轴的尺寸总是大于孔）；
(c) 过渡配合（限制条件可能是过盈配合的间隙）

图 1-7 所示为在装配图中配合尺寸标注的典型示例。

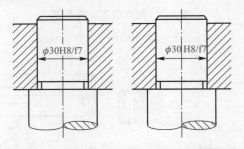

图 1-7 在装配图中配合尺寸标注的典型示例

1. 可互换性

可互换的零件是一种能由同一图纸加工出来的相似零件所代替的零件。零件的互换性基于以下两个功能。

（1）对于相互配合的零件来说，设计成同一尺寸公差是十分必要的。

（2）必须在指定的公差内制造零件。

2. 尺寸公差

决定一个具体尺寸公差时，有三个需要考虑的方面：功能重要性、互换性和经济性。

第一，有必要了解这个零件的用途；第二，了解失效场合下它的替换性；第三，要考虑避免在生产上花费不必要的时间和金钱。至于决定采用何种可行的公差等级，需要在精确性与经济性之间评判协调。为帮助设计者选择公差与配合，以及促进工业界的一致性（国内和国外），已经提出了许多公差与配合体系。

Evaluate

任务名称		工程制图	姓名	组别	班级	学号	日期	
		考核内容及评分标准		分值	自评	组评	师评	平均分
三维目标	知识	坐标系、视图类型、配合等概念		25分				
	技能	能阅读专业英语文章、翻译专用英语词汇		40分				
	素养	在学习过程中秉承科学精神、合作精神		35分				
加分项	收获（10分）	收获（借鉴、教训、改进等）：		你进步了吗？			加分	
				你帮助他人进步了吗？				
	问题（10分）	发现问题、分析问题、解决方法、创新之处等：					加分	
总结与反思							总分	

1.3 Computer Aided Design

Text

Computer aided design (CAD) refers to the use of computers and their graphic equipment to help designers carry out design work. It was born in the 1960s and was the interactive graphics research plan proposed by the United States Massachusetts Institute of Technology (MIT). Due to the expensive hardware facilities at that time, only the United States General Motors Company and the United States Boeing

Aviation Company used their own interactive mapping system.

CAD System's Work

Computer aided design systems are powerful tools and are used in the design and geometric modeling of products.

Drawings are generated at work stations, and the design is displayed continuously on the monitor in different colors for its various parts. The designer can easily conceptualize the part designed on the graphics screen and can consider alternative designs or modify a particular design to meet specific design requirements. Using powerful software such as computer aided three dimensional interactive applications (CATIA), the design can be subjected to engineering analysis and can identify potential problems, such as excessive load, deflection, or interference at mating surfaces during assembly. Information (such as a list of materials, specifications, and manufacturing instructions) is also stored in the CAD database. Using this information, the designer can analyze the manufacturing economics of alternatives.

Exchange Specifications

Because of the availability of a variety of CAD systems with different characteristics supplied by different vendors, effective communication and exchange of data between these systems are essential. Drawing exchange format (DEX) was developed for use with Autodesk and basically has become a standard because of the long term success of this software package. DEX is limited to transfer geometry information only. Similarly, stereo lithography (STL) formats are used to export three dimensional geometries, initially only to rapid prototyping system, but recently, it has become a format for data exchange between different CAD systems.

The necessity for a single, neutral format for better compatibility and for the transfer of more information than geometry alone is currently filled mainly by initial graphics exchange specification (IGES). This is used for translation in two directions (in and out of a system) and is also widely used for translation of three dimensional line and surface data. Because IGES is evolving, there are many variations of IGES.

Another useful format is a solid model based standard called the product data exchange specification (PDES), which is based on the standard for the exchange of product model data (STEP) developed by the International Standard Organization (ISO). PDES allows information on shape, design, manufacturing, quality assurance, testing, maintenance, etc., to be transferred between CAD systems.

Elements of CAD System

The design process in a CAD system consists of the four stages described as follows.

1. Geometric Modeling

In geometric modeling, a physical object or any of its parts is described mathematically or analytically. The designer first constructs a geometric model by giving commands that create or modify lines, surfaces, solids, dimensions, and text. Together, these propose an accurate and complete two or three dimensional representation of the object. The results of these commands are displayed and can be moved around on the screen, and any section desired can be magnified to view details. These data are stored in the database contained in computer memory.

The octree representation of a solid object is shown in Fig. 1-8. It is a three-dimensional analog to pixels on a television screen. Just as any area can be broken down into quadrants, any volume can be broken down into octants, which are then identified as solid, void, or partially filled. Partially filled voxels (from volume pixels) are broken into smaller octants and are reclassified. With increasing resolution, exceptional part detail is achieved. This process may appear to be some what cumbersome, but it allows for accurate description of complex surfaces. It is used particularly in biomedical applications, such as for modeling bone geometries.

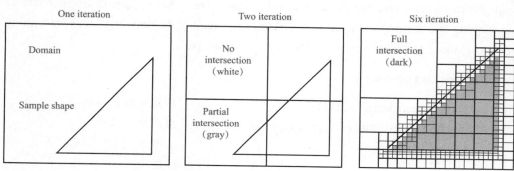

Fig. 1-8 The octree representation of a solid object

The octree representation of a solid object, any volume can be broken down into octants, which are then identified as solid, void or partially filled. Fig.1-8 is a two dimensional version (or quadtree) for the representation of shapes in a plane.

2. Design Analysis and Optimization

After the geometric features of a particular design have been determined, the design is subjected to an engineering analysis. This phase may consist of analyzing (for example) stresses, strains, deflections, vibrations, heat transfer, temperature distribution, or dimensional tolerances. Various sophisticated software packages are available, each having the capabilities to compute these quantities accurately and rapidly.

Because of the relative ease with which such analysis can now be done, designers increasingly are willing to analyze a design more thoroughly before it moves onto production. Experiments and measurements in the field nonetheless maybe necessary to determine the actual effects of loads, temperature, and other variables on the designed components.

3. Design Review and Evaluation

An important design stage is design review and evaluation used to check for any interference between various components. This is done in order to avoid difficulties during assembly or in use of the part and to determine whether moving members (such as linkages) are going to operate as intended. Software is available with animation capabilities to identify potential problems with moving members and other dynamic situations. During the design review and evaluation stage, the part is dimensioned and toleranced precisely to the full degree required for manufacturing it.

4. Documentation and Drafting

After the preceding stages have been completed, the design is reproduced by automated drafting machines for documentation and reference. At this stage, detailed and working views also are developed and printed. The CAD system is also capable of developing and drafting sectional views of the part, scaling the drawings, and performing transformations in order to present various views of the part.

【 Words and phrases 】

computer aided design (CAD)	计算机辅助设计
Massachusetts Institute of Technology (MIT)	麻省理工学院
conceptualize	vt. 使概念化
pixel	n. 像素
octant	n. 八分仪
tolerance	n. 公差
engineering analysis	工程分析
interactive computer graphic	交互式计算机绘图
octree representation	八叉树表示方法
drawing exchange format	图形交换格式
Autodesk	自动计算机辅助设计软件,如AutoCAD、3D Studio 等
software package	软件包
stereolithography	n. 光固化快速成型
geometric modeling	几何建模
quadtree	n. 四叉树
initial graphics exchange specification	初始图形交换标准

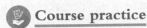 Course practice

1. Complex sentence analysis.

(1) The designer can easily conceptualize the part designed on the graphics screen

and can consider alternative designs or modify a particular design to meet specific design requirements.

　　designed on the graphics screen：作为 part 的后置定语。

　　【译文】设计人员可以不费力地在图形屏幕上将设计的零件概念化，并能考虑几种可能的设计方案，或者通过修正某种设计方案来满足具体的设计要求。

　　（2）Because of the availability of a variety of CAD systems with different characteristics supplied by different vendors, effective communication and exchange of data between these systems are essential.

　　【译文】由于不同经销商提供了各种不同特性的 CAD 系统，这些系统间的有效通信和数据交换是十分必要的。

　　（3）PDES allows information on shape, design, manufacturing, quality assurance, testing, maintenance, etc., to be transferred between CAD systems.

　　【译文】PDES 允许形状、设计、制造、质量保证、检测、维护等信息在 CAD 系统中传递。

　　（4）Just as any area can be broken down into quadrants, any volume can be broken down into octants, which are then identified as solid, void, or partially filled.

　　① Just as ...：正如……。

　　② which：引导一个非限定性定语从句，修饰前面的内容。

　　【译文】正如任何面积可以分解为四等份一样，任何体积也可以分解为八等份，然后作为实体、空白或颗粒填充进行识别。

　　（5）During the design review and evaluation stage, the part is dimensioned and toleranced precisely to the full degree required for manufacturing it.

　　① dimension：标注所需尺寸。

　　② tolerance：规定公差。

　　【译文】在设计评审阶段，零件被最大限度地标注准确尺寸并规定公差，以满足制造要求。

　　（6）The CAD system is also capable of developing and drafting sectional views of the part, scaling the drawings, and performing transformations in order to present various views of the part.

　　be capable of ...：能够……。

　　【译文】CAD 系统也能够生成零件剖面草图并进行缩放，同时也可以进行转换，以呈现零件的各种剖面。

　　2. After reading the text above summarize the main ideas in oral.

　　3. Fill in the blanks with proper words or phrases according to the text (note the proper tense).

　　（1）The CAD system is also c_____ developing and drafting sectional views of the part, s_____ the drawings, and p_____ transformations in order to present various views of the part.

（2）During the design review and evaluation stage, the part is d_____ and t_____ precisely to the full degree required for manufacturing it.

（3）After the geometric features of a particular design h_____, the design is subjected to an engineering analysis.

（4）Just as any area can be b_____ quadrants, any volume can be broken down into octants, which are then identified as solid, void, or partially filled.

（5）The designer can easily c_____ the part designed on the graphics screen and can consider a_____ designs or modify a particular design to meet specific design requirements.

4．Translate the following phrases into Chinese according to the text.

（1）CAD
（2）DEX
（3）IGES
（4）ISO
（5）be capable of
（6）be broken down into
（7）be based on
（8）consist of

5．Translate the following phrases into English according to the text.

（1）几何建模
（2）几何特征
（3）软件包
（4）设计评审
（5）设计优化
（6）八叉树表示方法

6．Write a 100-word summary for the text.

Translation of the text

计算机辅助设计

计算机辅助设计是指利用计算机及其图形设备帮助设计人员进行设计工作。其诞生于20世纪60年代，是美国麻省理工学院提出的交互式图形学的研究计划，由于当时硬件设施价格高，只有美国通用汽车公司和美国波音航空公司使用了自行开发的交互式绘图系统。

计算机辅助设计系统的工作

计算机辅助设计系统是强大的工具，它用于产品的设计与几何建模。

图纸在工作站生成，设计的不同零件以不同的颜色在显示器上连续显示。设计人员可以不费力地在图形屏幕上将设计的零件概念化，并能考虑几种可能的设计方案，或者通过修正某种设计方案来满足具体的设计要求。使用强大的软件如计算机辅助三维交互式应用（CATIA），能对设计进行工程分析和发现潜在的问题，如过载、偏斜或装配中配合面的过盈。那些信息（如材料清单、规格或加工说明书）也存储在CAD数据库中。使用这些信息，设计师可以分析几种加工方案的经济性。

交换标准

由于不同经销商提供了各种不同特性的 CAD 系统，这些系统间的有效通信和数据交换是十分必要的。由于软件包的长期成功应用，图形交换格式（DEX）被开发用于 Autodesk 且成为一个标准，DEX 仅局限于几何信息的转换。类似地，光固化快速成型（STL）格式被用于三维几何信息输出，其最初仅用于快速成型系统，但是最近它变成一种在不同 CAD 系统中进行数据交换的格式。

目前，主要由初始图形交换标准（IGES）满足对单一中性格式的需求，以实现更好的兼容性以及比单独几何传输更多的信息。其可用于两个方向的转换（系统的输入和输出），并且也广泛应用于三维直线或表面数据的转换。由于 IGES 不断升级，产生多种版本的 IGES。

另一种有用的格式是基于实体模型的标准，称为产品数据交换标准（PDES），它是基于国际标准化组织（ISO）开发的产品模型数据交换标准（STEP）。PDES 允许形状、设计、制造、质量保证、检测、维护等信息在 CAD 系统中传递。

CAD 系统的组成

CAD 系统中的设计过程包含以下四个阶段。

1. 几何建模

在几何建模中，一个实际物体或它的任意零件被数学或分析法描述。首先，设计师通过给出创建或修改直线、表面、实体、尺寸和文本等指定命令来建立几何模型，结合这些信息，对系统提出了物体准确、完整的二维或三维描述。这些命令的结果被显示且能在屏幕上移动，并且任何需要查看详细信息的部分可生成局部放大图。这些数据保存在计算机存储器的数据库中。

实体物体的八叉树表示方法如图 1-8 所示，它是三维的，类似电视屏幕上的像素。正如任何面积可以分解为四等份一样，任何体积也可以分解为八等份，然后作为实体、空白或颗粒填充进行识别。颗粒填充体素（来自空间体素）被划分成小的八等份且重新分类，随着不断更新，可获得特殊部件的细节。这个过程比较烦琐，但它可以精确描述复杂表面，特别适用于生物医学，如骨骼的几何建模。

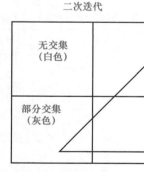
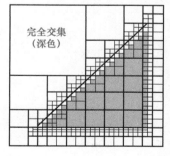

图 1-8 实体物体的八叉树表示方法

通过实体的八叉树表示方法可以知道，任何体积都可以被分解成八等份，然后被识别为固体、空白或局部填充体。图 1-8 中所展示的是平面形状表示的二维版本（或四叉树）。

2. 设计分析和优化

特定设计的几何特征被确定后，可对该设计进行工程分析。这个阶段包括对压力、应力、偏斜、振动、热传递、温度分布或尺寸公差等的分析。可使用各种复杂的软件包，每个软件包具有准确、快速计算这些物理量的能力。

由于完成分析相对容易，设计师越来越喜欢在实际生产前更全面地分析设计。在该领域在现场进行试验和检测是必要的，以确定负载、温度和其他变量对所设计零件的实际影响。

3. 设计评审与评估

用来检查各种零件间有无干扰的设计评审是一个重要的步骤。做设计评审是为了避免零件装配困难或使用困难，以及确定移动部件（如连杆）是否按预期的设计做运动。软件可以使用动画功能发现部件在运动和其他动态环境下的潜在问题。在设计评审阶段，零件被最大限度地标注准确尺寸并规定公差，以满足制造要求。

4. 文件存档与绘图

在完成上述阶段后，通过自动绘图仪重新生成设计，以便于存档和参考。在这一步骤中，局部放大图和工作图也被生成并打印。CAD系统也能够生成零件剖面草图并进行缩放，同时也可以进行转换，以呈现零件的各种剖面。

Evaluate

任务名称		计算机辅助设计	姓名	组别	班级	学号	日期
考核内容及评分标准			分值	自评	组评	师评	平均分
三维目标	知识	CAD制图的工作内容、交换标准、CAD系统的组成等	25分				
	技能	能阅读专业英语文章、翻译专用英语词汇	40分				
	素养	在学习过程中秉承科学精神、合作精神	35分				
加分项	收获（10分）	收获（借鉴、教训、改进等）：	你进步了吗？		加分		
			你帮助他人进步了吗？				
	问题（10分）	发现问题、分析问题、解决方法、创新之处等：			加分		
总结与反思					总分		

1.4 Mould Design and Manufacturing

Text

I. Introduction to Mould

Mould is a fundamental technological device for industrial production. Industrially produced goods are formed in moulds which are designed and built specially for them. The mould is the core part of manufacturing process because its cavity gives its shape. There are many kinds of moulds, such as casting and forging-dies, ceramic moulds, die-casting moulds, drawing dies, injection moulds, glass moulds, magnetic moulds, metal extruding moulds, plastic and rubber moulds, plastic extruding moulds, powder metallurgical moulds, compressing moulds, etc.

Compression Moulding

Compression moulding is the least complex of the moulding processes and is ideal for large parts or low-quantity production. For low-quantity requirements, it is more economical to build a compression mould than an injection mould. Compression moulds are often used for prototyping, where samples are needed for testing fit and forming into assemblies. This allows for further design modification before building an injection mould for high-quantity production. Compression moulding is best suited for designs where tight tolerances are not required (Fig. 1-9).

Injection Moulding

Injection moulding is the most complex of the moulding processes. Due to the more complex design of the injection mould, it is more expensive to purchase than a cast or compression mould. Although moulding cost can be high, cycle time is much faster than other processes and the part cost can be low, particularly when the process is automated. Injection moulding is well suited for moulding delicately shaped parts because high pressure (as much as 29, 000 psi, about 199.96 MPa) is maintained on the material to push it into every corner of the mould cavity (Fig. 1-10).

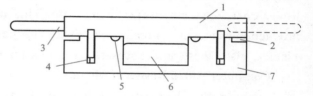

Fig. 1-9　A compression mould
1— top plate; 2— opening bar slot; 3— handle; 4— dowel pin and bushing; 5— flash and tear trim gate; 6— cavity; 7— bottom plate

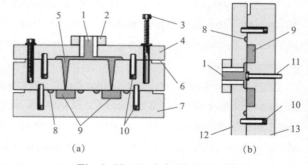

Fig. 1-10　An injection mould
(a) Horizontal type; (b) Vertical type
1— injection runner; 2— nozzle bushing; 3— stripper bolt; 4— top plate; 5— sprue; 6— opening bar slot; 7— bottom plate; 8— flash and tear trim gate; 9— cavity; 10— dowel pin and bushing; 11— ejector; 12— fixed plate; 13— movable plate

Cast Moulding

There are two types of casting, open casting and pressure casting. With open casting, the liquid mixture is poured into the open cavity in the mould and allowed to cure. With pressure casting, the liquid mixture is poured into the open cavity, the cap is put in place and the cavity is pressurized. Pressure casting is used for more complex parts and when moulding foam materials.

Extrusion Moulding

Although extrusion moulds are quite simple, the extrusion moulding process requires great care in the setting up and manufacturing and final processing to ensure consistency of the product. pressure is forced through the die plate that has the correct profile cut into it. Variations in feed rate, temperature and pressure need to be controlled.

II. Mould Design and Manufacturing

CAD and computer aided manufacturing (CAM) are widely applied in mould design and mould making. On one hand, CAD allows you to draw a model on the screen, then to view it from every angle using 3D animation, finally, to test it by introducing various parameters into the digital simulation models. CAM, on the other hand, allows you to control the manufacturing quality. The advantages of these computer technologies are legion: shorter design times (modifications can be made at the speed of the computer), lower cost, faster manufacturing, etc. This new approach also allows shorter production runs, and to make last-minute changes to the mould for a particular part. Finally, also, these new processes can be used to make complex parts.

Computer Aided Design of Moulds

Traditionally, the creation of drawings of moulds has been a time-consuming task that is not part of the creative process. Drawings are an organizational necessity rather than a desired part of the process.

CAD systems offer an efficient means of design, and can be used to create inspection programs when used in conjunction with coordinate measuring machines and other inspection equipment. CAD data can also play a critical role in selecting process sequence.

A CAD system consists of three basic components: hardware, software, users. The hardware components of a typical CAD system include a processor, a system display, a keyboard, a digitizer, and a plotter. The software components of a CAD system consists of the programs which allow it to perform design and drafting functions. The user is the tool designer who uses the hardware and software to perform the design process.

Detailing the functional components and adding the standard components to complete the mould, as shown in Fig. 1-11, are all happens in 3D. Moreover, the mould system provides functions for the checking, modifying and detailing of the part. Already in this early stage, drawings and bill of materials can be created automatically.

Through the use of 3D and the intelligent mould design system, typical 2D mistakes—such as a collision between cooling and components/cavities or the wrong position of a hole—can be eliminated at the beginning. At any stage a bill of materials and drawings can be created—allowing the material to be ordered on time and always having an actual document to discuss with the customer or a bid for a mould base manufacturer.

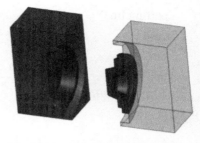

Fig. 1-11 3D solid model of mould

Computer Aided Manufacturing of Mould

One way to reduce the cost of manufacturing and reduce lead-time is by setting up a manufacturing system that uses equipment and personnel to their fullest potential. The foundation for this type of manufacturing system is the use of CAD data to help in making key process decisions that ultimately improve machining precision and reduce non-productive time. This is called as computer-aided manufacturing. The objective of CAM is to produce, if possible, sections of a mould without intermediate steps by initiating machining operations from the computer workstation.

【 Words and phrases 】

I

mould	n. 模具；模（型）；模塑；压模
cavity	n. 模腔；型腔；孔洞
die	n. 模（子/片/具）；压模；塑模
casting and forging die	铸锻模具
ceramic mould	陶瓷模具
die-casting mould	压铸模具
drawing die	拉丝模具
injection mould	注塑模具
magnetic mould	磁性模具
extruding mould	挤压模具
powder metallurgical mould	粉末冶金模具
compressing mould	冲压模具
moulding	n. 成型
prototyping	n. 初始制模
assembly	n. 装配件；组件
tolerance	n. 公差；容许
ejector	n. 脱模销；推顶器
runner	n. 浇道；流道
sprue	n. 铸口

parting line	合模线；拼缝线
cure	v. 固化；塑化
pin	n. 销；杆
psi	abbr. 磅每平方英寸（pounds per square inch）

II

computer aided manufacturing（CAM）	计算机辅助制造
animation	n. 动画
digital simulation model	数字模拟模型
legion	n. 多；大批；无数
collision	n. 打击；碰撞
lead-time	n. 研制周期
bill of material	物料清单

Course practice

1. Complex sentence analysis.

（1）Although moulding cost can be high, cycle time is much faster than other processes and the part cost can be low, particularly when the process is automated.

① moulding cost：模具成本。

② particularly when the process is automated：条件状语从句放在后面，用于进一步说明，译为尤其当工艺过程为自动操作时。

【译文】虽然模具成本可能很高，但其制造周期比其他工艺短，零件成本比较低，尤其当工艺过程为自动操作时。

（2）Injection moulding is well suited for moulding delicately shaped parts because high pressure (as much as 29,000 psi) is maintained on the material to push it into every corner of the mould cavity...

① is well suited for：非常适合。

② delicately shaped parts：外形精致的零件。

③ psi：英制压力单位，磅每平方英寸。

【译文】注塑成型非常适合造型精致的零件，因为材料会被高压（高达29 000磅每平方英寸）推入模腔的每个角落……

（3）Although extrusion moulds are quite simple, the extrusion moulding process requires great care in the setting up and manufacturing and final processing to ensure consistency of the product.

① in the setting up and manufacturing and final processing：在设置、制造和最终加工过程中。

② to ensure consistency of the product：译为以确保产品的一致性，表示目的。

【译文】虽然挤压模具相对简单,但挤压成型过程在设置和制造,以及最终加工中需要非常小心,以确保产品的一致性。

(4) CAD allows you to draw a model on the screen, then to view it from every angle using 3D animation, finally, to test it by introducing various parameters into the digital simulation models.

① draw、view 和 test 表示 3 个并列的操作。

② digital simulation model:数字仿真模型。

【译文】CAD 允许在屏幕上绘制模型,然后使用 3D 动画从各个角度查看它,最后通过将各种参数引入数字仿真模型来测试它。

(5) At any stage a bill of materials and drawings can be created—allowing the material to be ordered on time and always having an actual document to discuss with the customer or a bid for a mould base manufacturer.

① a bill of materials:材料清单。

② allowing 和 having 引导两个现在分词短语,作伴随状语。

【译文】在任何阶段都可以创建材料清单和图纸,以便按时订购材料,并始终有实际的文件与客户讨论或为模具基础制造商投标。

(6) The objective of CAM is to produce, if possible, sections of a mould without intermediate steps by initiating machining operations from the computer workstation.

① if possible:插入语,译为如果可能的话,是 if it is possible 的省略语。试比较:

If possible, I will visit you in Chicago next month.

I will lend you some money to help you with the present difficulty if possible.

② by...:通过……方式。

【译文】CAM 的目标是生产制造,如果可能的话,部分模具在计算机工作站就可以开始加工,而省略中间环节。

2. Translate the following paragraphs.

(1) The most common types of moulds used in industry today are two-plate moulds, three-plate moulds, side-action moulds, and unscrewing moulds.

(2) A two-plate mould consists of two active plates, into which the cavity and core inserts are mounted. In this mould type, the runner system, sprue, runners, and gates solidify with the part being moulded and are ejected as a single connected item. Thus the operation of a two-plate mould usually requires continuous machine attendance.

(3) A key decision early in mold making process is determining what machining operations will be used and in what order. Machining considerations should be analyzed during the development of the CAD model. If this isn't done, the programmer may not be able to use certain machining strategies.

(4) Each of the processes has advantages and disadvantages when producing a close tolerance mould. The proper selection of process and sequence of process will not only

result in more precise dimensional control, but also will reduce manufacturing time by reducing bench work.

3. Choose the proper answer to fill in the blank and translate the sentences.

(for, of, at, against, through)

(1) Industrially produced goods are formed in moulds which are designed and built specially (   ) them.

(2) Due to the more complex design (   ) the injection mould, it is more expensive to purchase than a cast or compression mould.

(3) Injection moulding is well suited (   ) moulding delicately shaped parts.

(4) At any stage a bill (   ) materials and drawings can be created.

Translation of the text

模具设计和制造

I. 模具概论

模具是工业生产的基本工艺装置，工业产品是在为其设计和制造的专用模具中成型的。模具是制造过程的核心部分，因为它的型腔决定了产品的形状。模具有很多种，如铸锻模具、陶瓷模具、压铸模具、拉丝模具、注塑模具、玻璃模具、磁性模具、金属挤压模具、塑料橡胶模具、塑料挤压模具、粉末冶金模具、压缩模具等。

压缩成型

压缩成型是最不复杂的成型工艺，适用于大零件或小批量生产。对于低数量的需求，创建一个压缩模具比创建一个注塑模具更经济。压缩模具通常用于原型制作，需要样品来测试配合并形成组件，这样可以在创建用于大批量生产的注塑模具之前进行进一步的设计修改。压缩成型最适合不需要严格公差的产品设计。图1-9所示为压缩模具结构。

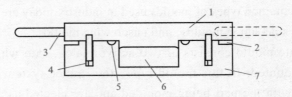

图1-9 压缩模具结构
1—顶板；2—开杆槽；3—手柄；4—定位销和衬套；5—闪撕闸板；6—腔；7—底板

注塑成型

注塑成型是最复杂的成型工艺。由于注塑模具的设计比较复杂，比铸造模具或压缩模具价格更高。虽然模具成本可能很高，但其制造周期比其他工艺短，零件成本比较低，尤其当工艺过程为自动操作时。注塑成型非常适合造型精致的零件，因为材料会被高压（高达29 000 psi，约为199.96 MPa）推入模腔的每个角落。图1-10所示为注塑模具结构。

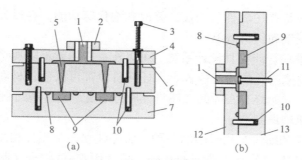

图 1-10 注塑模具结构
(a) 水平式；(b) 垂直式

1—注射流道；2—喷嘴套管；3—脱模螺钉；4—顶板；5—浇口；6—开杆槽；7—底板；
8—闪撕闸板；9—腔；10—定位销和衬套；11—喷射器；12—固定板；13—可移动板

铸造成型

铸造有两种类型，即开口铸造和压力铸造。开口铸造，即将液体混合物倒入模具的开口腔中并固化。压力铸造，即将液体混合物倒入开口的腔内，盖上盖子，对腔加压。压力铸造用于更复杂的零件和泡沫材料成型。

挤压成型

虽然挤压模具相对简单，但挤压成型过程在创设和制造，以及最终加工中需要非常小心，以确保产品的一致性。将压力施加于模板，按正确的轮廓进行切割，需要控制进料速度、温度和压力的变化。

Ⅱ. 模具设计和制造

CAD 和 CAM 在模具设计和模具制造中有广泛的应用。一方面，CAD 允许在屏幕上绘制模型，然后使用 3D 动画从各个角度查看它，最后通过将各种参数引入数字仿真模型来测试它。另一方面，CAM 允许控制制造质量。计算机技术的优点很多：更短的设计时间（以计算机的速度进行修改）、更低的成本、更快的制造速度等。这种新方法还可以缩短生产周期，可以在最后一刻改变模具的特定部分。最后，这些新工艺也可以用来制造复杂的零件。

模具的计算机辅助设计

传统上，模具图纸的创建是一项耗时的任务，不是创作过程的一部分。图纸是筹备的必需品，而不是过程中期望的一部分。

CAD 系统提供了一种有效的设计手段，与坐标测量机和其他检测设备一起使用时，可以用来创建检测程序。CAD 数据在选择工艺顺序方面也起着至关重要的作用。

一个 CAD 系统由三个基本部分组成：硬件、软件和用户。典型 CAD 系统的硬件组件包括处理器、系统显示器、键盘、数字转换器和绘图仪。CAD 系统的软件由可执行设计和绘图功能的程序组成。用户是使用硬件和软件来执行设计过程的工具设计者。

对功能部件进行细化，添加标准件完成模具，（如图 1-11 所示），这一切都发生在三维空间中。

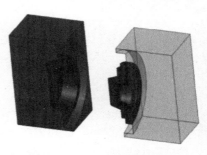

图 1-11 模具的三维实体模型

另外，模具系统还提供了零件的校核、修改和细化等功能。在设计的早期阶段，已经可以自动创建图纸和材料清单。

通过使用三维和智能模具设计系统，可以在一开始就消除典型的二维错误，如冷却与部件/腔碰撞或孔的错误位置。在任何阶段都可以创建材料清单和图纸，以便按时订购材料，并始终有实际的文件与客户讨论或为模具基础制造商投标。

模具的计算机辅助制造

降低制造成本和缩短交货时间的一种方法是建立一个充分利用设备和人员潜力的制造系统。这种类型的制造系统的基础，是使用CAD数据来帮助制定关键的工艺决策，最终提高加工精度并减少非生产时间。这被称为计算机辅助制造。（CAM的目标是生产制造，如果可能的话，部分模具在计算工作站就可以开始加工，而省略中间环节。）

Evaluate

任务名称		模具设计和制造		姓名	组别	班级	学号	日期	
考核内容及评分标准					分值	自评	组评	师评	平均分
三维目标	知识	压缩成型、注塑成型、铸造成型等概念			25分				
	技能	能阅读专业英语文章、翻译专用英语词汇			40分				
	素养	在学习过程中秉承科学精神、合作精神			35分				
加分项	收获（10分）	收获（借鉴、教训、改进等）：			你进步了吗？			加分	
					你帮助他人进步了吗？				
	问题（10分）	发现问题、分析问题、解决方法、创新之处等：						加分	
总结与反思								总分	

Unit 2 Machine Manufacturing

Unit objectives

【Knowledge goals】
1. 掌握冷成型和热成型、塑性成型的原理，以及材料成型的方法等。
2. 掌握切削速度、进给运动、切割深度等概念。
3. 掌握逆铣、顺铣等概念。
4. 掌握热处理曲线转化、退火、正火和球化等概念。
5. 掌握虚拟制造技术的概念、分类、虚拟机等相关知识。

【Skill goals】
1. 能够阅读专业英语文章。
2. 可以翻译专用英语词汇。

【Quality goals】
1. 在学习过程中发扬科学研究精神。
2. 增强团队合作意识。

2.1 Introduction to Material Forming

Text

I. Material Forming Processes

The term material forming refers to a kind of manufacturing methods by which the given shape of a workpiece (a solid body) is converted to another shape without change in the mass or composition of the material of the workpiece.

Cold Forming and Hot Forming

By applying a stress that exceeds the original yield strength of metallic material, we have strain hardened or cold worked the metallic material, while simultaneously deforming it. This is the basis for many manufacturing techniques, such as wire drawing. Fig. 1-12 illustrates several manufacturing processes that make use of both cold-working and hot-working processes.

Principles of Plastic Forming

Plasticity theory is the foundation for the numerical treatment of metal forming processes. Materials science and metallurgy can explain the origins of the plastic state of metallic bodies and its dependence on various parameters, such as process speed, prior history, temperature, and so on. The essentially older plasticity theory deals with the calculation of stresses, forces, and deformation.

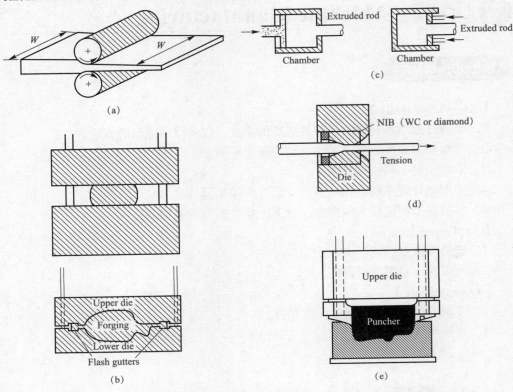

Fig. 1-12 Manufacturing processes that make use of cold working and hot working
(a) Rolling; (b) Forging (open and chosen die); (c) Extrusion; (d) Wire drawing; (e) Stamping

Methods Used in Material Forming

The following classification of the deformation methods is based mainly on the important differences in effective stresses. No simple descriptions of stress states are possible, since depending on the kind of operation, different stress states may occur simultaneously, or they may change during the course of the deforming operation. Therefore, the predominant stresses are chosen as the classification criteria.

II. Material Forming Examples

In this section, a short description of the process examples about material forming will be given. But assembly and joining processes are not described here.

Forging

Forging can be characterized as mass conservation, solid state of workpiece material (metal), and mechanical primary basic process—plastic deformation. Technically, forging may be defined as the process of giving metal increased utility by shaping it, refining it, and improving its mechanical properties through controlled plastic deformation under impact or pressure. Fig.1-13 (a) shows the drop forging process.

Rolling

Rolling can be characterized as mass conservation, solid state of workpiece material, mechanical primary basic process—plastic deformation. Rolling is extensively used in the manufacturing of plates, sheets, structural beams, and so on. Fig.1-13 (b) shows the rolling process of a plate. An ingot is produced in casting and in several stages it is reduced in thickness, usually while hot. Since the width of the workpiece material is kept constant, its length is increased according to the reductions. After the last hot rolling stage, the final stage of cold rolling is carried out to improve surface quality and tolerances and to increase strength. In rolling, the profiles of the rolls are designed to produce the desired geometry.

Powder Compaction

Powder compaction can be characterized as mass conservation, granular state of material, mechanical basic process—flow and plastic deformation. In this context, only compaction of metal powder is mentioned [Fig.1-13 (c)], but generally compaction of molding sand, ceramic materials, and so on, also belong in this category.

Casting

Casting can be characterized as mass conservation, fluid state of material, mechanical basic process—filling of the die cavity. Casting is one of the oldest manufacturing methods and one of the best known processes in which the material is melted and poured into a die cavity corresponding to the desired geometry to form the shape [Fig.1-13 (d)].

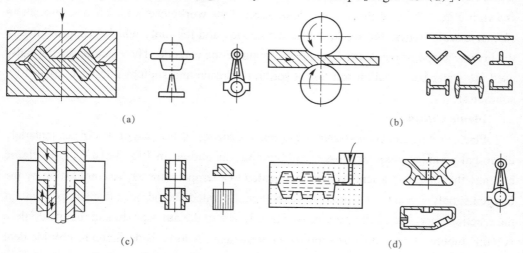

Fig. 1-13 Mass-conserving processes in the solid state of the workpiece material
(a) Drop forging; (b) Rolling of a plate; (c) Compaction of metal powder; (d) Casting

Turning

Turning can be characterized as mass reducing, solid state of workpiece material, mechanical primary basic process—fracture. The turning process, which is the best known and most widely used mass-reducing process, is employed to manufacture all types of cylindrical shapes by removing material in the form of chips from the workpiece material with a cutting tool, as shown in Fig. 1-14 (a). The workpiece rotates and the cutting tool is fed longitudinally. The cutting tool is much harder and more wear resistant than the workpiece material. A variety of types of lathes are employed, some of which are automatic in operation. The lathes are usually powered by electric motors which, through various gears, supply the necessary torque to the workpiece and provide the feed motion to the cutting tool.

Electrical Discharge Machining

Electrical discharge machining (EDM) can be characterized as mass reducing, solid state of workpiece material, thermal primary basic process—melting and evaporation, as shown in Fig. 1-14 (d). In EDM, material is removed by the erosive action of numerous small electrical discharges (sparks) between the work piece material and the tool (electrode), the latter having the inverse shape of the desired geometry. Each discharge occurs when the potential difference between the workpiece and the tool is large enough to cause a breakdown in the fluid medium, which is fed into the gap between the tool and workpiece under pressure, producing a conductive spark channel. The fluid medium, which is normally mineral oil or kerosene, has several functions.

Electrochemical Machining

Electrochemical machining (ECM) can be characterized as mass reducing, solid state of workpiece material, chemical primary basic process—electrolytic dissolution, as shown in Fig. 1-14 (e). Electrolytic dissolution of the workpiece is established through an electric circuit, where the workpiece is the anode, and the tool, which is approximately the inverse shape of the desired geometry, is made the cathode. The electrolytes normally used are water based saline solutions (sodium chloride and sodium nitrate in 10%-30% solutions).

Flame Cutting

Flame cutting can be characterized as mass reducing, solid state of workpiece material, chemical primary basic process—combustion, as shown in Fig. 1-14 (f). In flame cutting, the material (a ferrous metal) is heated to a temperature where combustion by the oxygen supply can start. Theoretically, the heat liberated should be sufficient to maintain the reaction once started, but because of heat losses to the atmosphere and the material, a certain amount of heat must be supplied continuously. A torch is designed to provide heat both for starting and maintaining the reaction. The most widely used is the oxyacetylene cutting torch, where heat is created by the combustion of acetylene and oxygen.

Fine Blanking

Fine blanking is a technique used for production of blanks perfectly flat and with a cut edge which is comparable to a machined finish. This quick and easy process is worthy of serious thought when the number of parts justifies the cost of a blanking tool especially when consideration is given to the fact that operations such as shaving are eliminated.

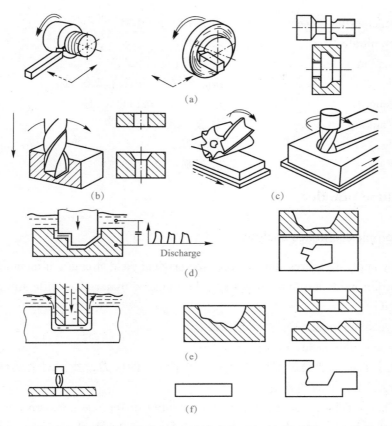

Fig. 1-14 Mass-reducing processes in the solid state of the workpiece
(a) Turning; (b) Drilling; (c) Milling; (d) Electrical discharge machining;
(e) Electrochemical machining; (f) Flame cutting

【Words and phrases】

I

workpiece	n. 工件；轧件；工件壁厚
metallurgy	n. 冶金；冶金学；冶金术
predominant	adj. 支配的；主要的；突出的

II

ingot	n. [冶] 锭铁；工业纯铁
profile	n. 剖面；侧面；外形；轮廓
ceramic	adj. 陶器的
longitudinally	adv. 经度上；纵向地
electrical discharge machine (EDM)	n. 电火花加工机床
erosive	adj. 侵蚀性的；腐蚀性的
dissolution	n. 分解；解散
electrolyte	n. 电解；电解液

sodium chloride	氯化钠
sodium nitrate	硝酸钠
combustion	n. 燃烧；燃烧过程
ferrous	adj. 含铁的；［化］亚铁的
acetylene	n. ［化］乙炔；电石气
fine blanking	精密冲裁
cut edge	剪切刃
shaving	n. 刨削；修整

Course practice

1. Complex sentence analysis.

（1）By applying a stress that exceeds the original yield strength of metallic material, we have strain hardened or cold worked the metallic material, while simultaneously deforming it.

① have strain hardened：产生形变硬化。

② while simultaneously：而同时。

【译文】通过施加超过金属材料原始屈服强度的应力，我们对金属材料进行应变硬化或冷加工，同时使其变形。

（2）Each discharge occurs when the potential difference between the workpiece and the tool is large enough to cause a breakdown in the fluid medium, which is fed into the gap between the tool and workpiece under pressure, producing a conductive spark channel.

① when：引导时间状语从句，修饰 occurs。

② ...fed into...：定语从句，修饰 fluid medium。

【译文】当工件和工具之间的电位差大到足以使流体介质被击穿时，就会发生放电，流体介质在压力下进入工具和工件之间的间隙，产生导电火花通道。

2. Translate the following paragraphs.

（1）Ceramic tools are a post-war introduction and are not yet in general factory use. Their most likely applications are in cutting metal at very high speeds, beyond the limits possible with carbide tools. Ceramics resist the formation of a built-up edge and in consequence produce good surface finishes.

（2）Since the present generation of machine tools is designed with only sufficient power to exploit carbide tooling, it is likely that, for the time being, ceramics will be restricted to high-speed finish machining where there is sufficient power available for the light cuts taken. The extreme brittleness of ceramic tools has largely limited their use to continuous cuts, although their use in milling is now possible.

3. Choose the proper answer to fill in the blank and translate the sentences.

(for, as, into, to, through, against, from)

(1) Plasticity theory is the foundation (　　) the numerical treatment of metal forming processes.

(2) The predominant stresses are chosen (　　) the classification criteria.

(3) The material is melted and poured (　　) a die cavity corresponding to the desired geometry.

(4) Theoretically, the heat liberated should be sufficient (　　) maintain the reaction once started.

Translation of the text

材料成型概论

I. 材料成型工艺

材料成型作为专业术语是指在不改变工件材料的质量或成分的情况下，将工件（实体）的给定形状转换为另一种形状的一种制造方法。

冷成型和热成型

通过施加超过金属材料原始屈服强度的应力，我们对金属材料进行应变硬化或冷加工，同时使其变形。这是许多制造技术的基础，如拉丝。图1-12所示为几种同时利用冷加工和热加工的成型工艺。

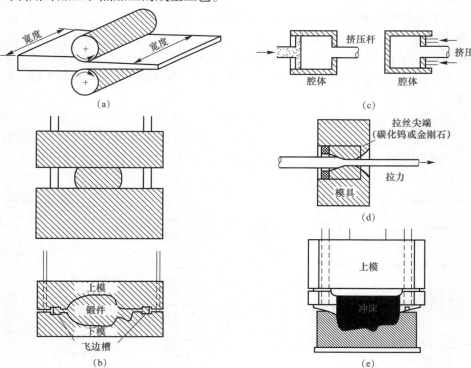

图 1-12　利用冷加工和热加工的成型工艺
(a) 滚轧；(b) 锻造（开模和选模）；(c) 挤压；(d) 拉丝；(e) 冲压

塑性成型原理

塑性理论是金属成型过程数值处理的基础。材料科学和冶金学可以解释金属体塑性状态的起源及其各种参数，如工艺速度、既往史、温度等。实际上旧的塑性理论处理的是应力、力和变形的计算。

材料成型的方法

对变形方法分类主要是基于有效应力的重要差异。不可能简单地描述应力状态，根据不同的操作类型，不同的应力状态可能同时发生，或者它们可能在变形操作过程中发生变化。因此，选择主要应力作为分类标准。

II. 材料成型工艺示例

在本部分，将简要描述有关材料成型工艺的示例，但不描述装配和连接过程。

锻造

锻造的特点：质量守恒、工件材料（金属）呈固态、机械初级基本工艺——塑性变形。从技术上讲，锻造可以定义为通过控制冲击或压力下材料的塑性变形，使金属成型、精炼并改善其力学性能，从而增加金属效用的过程。图1-13（a）所示为模锻法。

轧制

轧制的特点：质量守恒、工件材料呈固态、机械初级基本工艺——塑性变形。轧制广泛应用于板材、薄板、结构梁等的制造。图1-13（b）所示为板材的轧制过程。铸锭是在铸造过程中生产出来的，在几个阶段中，铸锭的厚度通常是在热的时候减小的。由于工件材料的宽度保持不变，因此其长度根据减少的量而增加。在热轧阶段之后，进行最后的冷轧阶段，以改善表面质量和公差，并提高强度。在轧制过程中，轧辊的轮廓被设计成所需的几何形状。

粉末压实

粉末压实的特点：质量守恒、材料呈粒状、机械基本工艺——流动和塑性变形。在此只提到了金属粉末的压实［图1-13（c）］，但一般型砂、陶瓷材料等的压实也属于这一类。

铸造

铸造的特点：质量守恒、材料呈流体状态、机械基本工艺——型腔填充。铸造是最古老的制造方法之一，也是最著名的工艺之一，在铸造过程中将材料熔化并倒入与所需几何形状对应的模腔中使材料成型，如图1-13（d）所示。

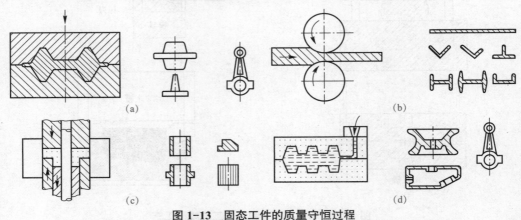

图1-13　固态工件的质量守恒过程

(a) 模锻法；(b) 板材的轧制；(c) 金属粉末的压实；(d) 铸造

车削

车削的特点：质量还原、工件材料呈固态、机械初级基本工艺——断裂。车削加工是目前最著名和应用最广泛的减重工艺，通过切削刀具从工件上去除材料来制造各种类型的圆柱形，如图 1-14（a）所示。工件旋转，刀具纵向进给。刀具比工件材料更硬、更耐磨。使用的各种类型的车床中，一些是自动操作的。车床通常由电动机驱动，电动机通过各种齿轮为工件提供必要的扭矩，并为刀具提供进给运动。

电火花加工

电火花加工（EDM）的特点：质量还原、工件材料呈固态、热初级基本工艺——熔化和蒸发，如图 1-14（d）所示。在电火花加工中，通过工件材料和工具（电极）之间的许多微放电（火花）的侵蚀作用去除材料，后者（工具电极）具有所需几何形状的相反形状。当工件和工具之间的电位差大到足以使流体介质被击穿时，就会发生放电，流体介质在压力下进入工具和工件之间的间隙，产生导电火花通道。流体介质通常是矿物油或煤油，具有多种功能。

电化学加工

电化学加工（ECM）的特点：质量还原、工件材料呈固态、化学初级基本工艺——电解溶解，如图 1-14（e）所示。通过电路建立工件的电解溶解，其中工件作为阳极，而将与所需几何形状近似相反的工具作为阴极。通常使用的电解质是水基生理盐水溶液（10%～30% 的氯化钠和硝酸钠溶液）。

火焰切割

火焰切割的特点：质量还原、工件材料呈固态、化学初级基本工艺——燃烧，如图 1-14（f）所示。在火焰切割中，材料（含铁金属）被加热到氧气供应时可以开始燃烧的温度。从理论上说，释放的热量应该足以维持反应开始，但由于热量损失到空气和材料中，必须持续提供一定量的热量。火炬的作用是为开始和维持反应提供热量。最广泛使用的是氧乙炔切割炬，它的热量是由乙炔和氧气燃烧产生的。

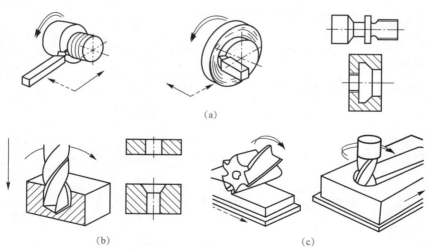

图 1-14　固态工件材料质量还原过程
(a) 车削；(b) 钻削；(c) 铣削

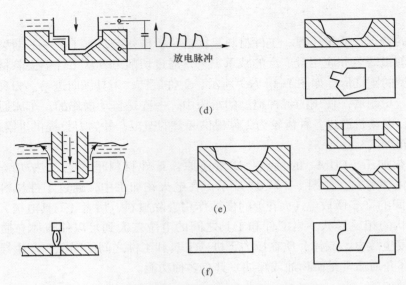

图 1-14 固态工件材料质量还原过程（续）
(d) 电火花加工；(e) 电化学加工；(f) 火焰切割

精密冲裁

精密冲裁是一种用于生产完全平整的毛坯的技术，具有可与机械加工相媲美的切割边缘。当零件数量超过了冲裁工具的成本时，特别是考虑到当切削之类的操作被限制时，可以考虑采用这种快速、简便的工艺。

Evaluate

任务名称		材料成型概论	姓名	组别	班级	学号	日期
考核内容及评分标准			分值	自评	组评	师评	平均分
三维目标	知识	冷成型和热成型、塑性成型的原理，材料成型方法等	25分				
	技能	能阅读专业英语文章、翻译专用英语词汇	40分				
	素养	在学习过程中秉承科学精神、合作精神	35分				
加分项	收获（10分）	收获（借鉴、教训、改进等）：	你进步了吗？			加分	
			你帮助他人进步了吗？				
	问题（10分）	发现问题、分析问题、解决方法、创新之处等：				加分	
总结与反思						总分	

2.2 Metal Cutting Processes

Text

Metal-cutting processes are extensively used in the manufacturing industry. They are characterized by the fact that the size of the original workpiece is sufficiently large that the final geometry can be circumscribed by it, and that the unwanted material is removed as chips, particles, and so on. The chips are a necessary means to obtain the desired tolerances, and surface roughness. The amount of scrap may vary from a few percent to 80% of the volume of the original workpiece.

Owing to the rather poor material utilization of the metal-cutting processes, the anticipated scarcity of materials and energy, and increasing costs, the development in the last decade has been directed toward an increasing application of metal-forming processes. However, die costs and the capital cost of machines remain rather high; Consequently, metal-cutting processes are, in many cases, the most economical, in spite of the high material waste, which only has value as scrap. Therefore, it must be expected that the material removal processes will for the next few years maintain their important position in manufacturing. Furthermore, the development of automated production systems has progressed more rapidly for metal-cutting processes than for metal-forming processes.

Metal-cutting processes remove material from the surface of a workpiece by producing chips. This requires that the cutting tool material be harder than the workpiece material. The final geometry of the workpiece is thus determined from the geometry of the tool and the pattern of motions of the tool and the workpiece. The basic process is mechanical, actually a shearing action combined with fracture.

As mentioned previously, the unwanted material in metal-cutting processes is removed by a rigid cutting tool, so that the desired geometry, tolerances, and surface roughness are obtained. Examples of processes in this group are turning, drilling, reaming, milling, shaping, broaching, grinding, honing, and lapping.

Most of the cutting or machining processes are based on a two dimensional surface creation, which means that two relative motions are necessary between the cutting tool and the workpiece. These motions are defined as the primary motion, which mainly determines the cutting speed, and the feed motion, which provides the cutting zone with new material.

In turning, the primary motion is provided by the rotation of the workpiece, and in planing it is provided by the translation of the table; in turning, feed motion is a continuous translation of the tool, and in planing it is an intermittent translation of the tool. Fig. 1-15 shows

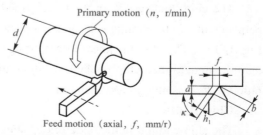

Fig. 1-15 Basic machining parameters in turning

basic machining parameters in turning.

Cutting Speed

The cutting speed v is the instantaneous velocity of the primary motion of the tool relative to the workpiece (at a selected point on the cutting edge).

The cutting speed for turning, drilling, and milling processes can be expressed as

$$v = \pi d n \quad (1-1)$$

where v is the cutting speed, measured in m/min; d is the diameter of the workpiece to be cut, measured in m; and n is the workpiece or spindle rotation measured in r/min. Thus v, d and n may relate to the workpiece or the tool, depending on the specific kinematic pattern. In grinding the cutting speed is normally measured in m/s.

Feed

The feed f is provided to the cutting tool or the workpiece, when added to the primary motion, leads to a repeated or continuous chip removal and the creation of the desired machined surface. The motion may proceed by steps or continuously. The feed speed v_f is defined as the instantaneous velocity of die feed motion relative to the workpiece (at a selected point on the cutting edge).

For turning and drilling, the feed f is measured in milimeter per revolution (mm/r) of the workpiece or the tool; for planing and shaping f is measured in milimeter per stroke (mm/stroke) of the tool or the workpiece. In milling the feed f_z is measured in milimeter per tooth of the cutting tool (mm/tooth); that is, f_z is the displacement of the workpiece between the cutting action of two successive tooth shaping. The feed speed v_f (mm/min) of the table is therefore the product of the number of teeth z of the cutting tool, the revolutions per minute of the cutter n, and the feed per tooth ($v_f = nzf_z$).

A plane containing the directions of the primary motion and the feed motion is defined as the working plane, since it contains the motions responsible for the cutting action.

Depth of Cut

In turning (Fig.1-15) the depth of cut a (also called back engagement) is the distance that the cutting edge engages or projects below the original surface of the workpiece. The depth of cut determines the final dimensions of the workpiece. In turning, with an axial feed, the depth of cut is a direct measure of the decrease in radius of the workpiece and with radial feed the depth of cut is equal to the decrease in the length of workpiece.

Chip Thickness

The chip thickness h_1 in the undeformed state is the thickness of the chip measured perpendicular to the cutting edge and in a plane perpendicular to the direction of cutting (see Fig. 1-15). The chip thickness after cutting (i.e., the actual chip thickness h_2) is larger than the undeformed chip thickness, which means that the cutting ratio or chip thickness ratio $r = h_1/h_2$ is always less than unity.

Chip Width

The chip width b in the undeformed state is the width of the chip measured along the cutting edge in a plane perpendicular to the direction of cutting.

Area of Cut

For single-point tool operations, the area of cut A is the product of the undeformed chip thickness h_1 and the chip width b (i.e., $A=h_1 b$). The area of cut can also be expressed by the feed f and the depth of cut a as follows:

$$h_1 = f \sin\kappa$$
$$b = a/\sin\kappa \qquad (1-2)$$

where κ is the tool cutting edge angle.

Consequently, the area of cut is given by

$$A = fa \qquad (1-3)$$

【 Words and phrases 】

workpiece	n. 工件；轧件；工件壁厚
circumscribe	vt. 约束，限定；在……上画圈；包围；（几何学）外接
chip	n. 芯片；碎块；碎屑；缺口
	v. 打破；弄缺；（尤指用工具）削下
scrap	n. 碎片；切屑；废品；废渣
	v. 废弃；使成为碎屑
scarcity	n. 不足；缺乏
primary motion	主运动
feed motion	进给运动
intermittent	adj. 间歇的；断断续续的
instantaneous	adj. 瞬间发生的；瞬间完成的
stroke	v. 打击；冲程；行程；循环；周期
feed per tooth	每齿进给量
engagement	n. 啮合
back engagement	背吃刀量
engage	v. 啮合；切入；雇用；聘请；参加；从事；吸引
unity	n. 团结；统一；整体性；统一性；[数]一
tool cutting edge angle	主偏角

Course practice

1. Complex sentence analysis.

For turning and drilling, the feed f is measured per revolution (mm/r) of the workpiece or the tool...

① is measured：被测量。

② ...per revolution：每转……

【译文】对于车削和钻孔，进给量 f 是按照工件或刀具每转一圈（mm/r）来测量的……

2．Translate the following sentence.

The chip width *b* in the undeformed state is the width of the chip measured along the cutting edge in a plane perpendicular to the direction of cutting.

3．Choose the proper answer to fill in the blank and translate the sentences.

（to，of，from，through，against，for）

（1）The amount of scrap may vary from a few percent（　　）80% of the volume of the original work material.

（2）In many cases, the most economical, in spite（　　）the high material waste, which only has value as scrap.

（3）Metal-cutting processes remove material（　　）the surface of a workpiece by producing chips.

 Translation of the text

金属切削工艺

金属切削工艺广泛应用于制造业。其特点是原始工件的尺寸足够大，最终的几何形状可以由刀具的几何形状来限定，并且将不需要的材料形成切屑、颗粒等去除。切屑是获得所需公差和表面粗糙度的必要手段。废料的数量是原来工件体积的百分之几到80%不等。

由于金属切削工艺的材料利用率较低、预计材料和能源的稀缺及成本的增加，过去10年的发展方向是金属成型工艺的应用不断增加。然而，模具成本和机器的成本仍然很高；因此，在许多情况下，虽然金属切削工艺浪费很多材料，这些废弃材料只有废料的价值，但还是最经济的。因此，可以预料材料去除工艺在未来几年内将保持在制造业中的重要地位。另外，自动化生产系统的发展在金属切削工艺中，比在金属成型工艺中发展得更快。

金属切削工艺通过制造切屑从工件表面去除材料，这就要求切削刀具材料比工件材料硬。因此，工件的最终几何形状是由刀具的几何形状，以及刀具和工件的运动模式确定的。其基本工艺是机械的，实际上是剪切作用与断裂相结合。

如前所述，金属切削工艺中不需要的材料被刚性切削刀具去除，从而获得所需要的几何形状、公差和表面粗糙度。这类加工的例子有车削、钻孔、铰孔、铣削、成型、拉削、磨削、珩磨、研磨。

大多数切削或加工工艺都是基于二维表面的创建，这意味着切削刀具和工件之间需要进行两次相对运动。这些运动被定义为主要决定切削速度的初级运动，以及为切削区提供新材料的进给运动。

在车削工艺中，主运动是由工件的旋转提供的，在刨削时主运动是由工作台的平移提供的；在车削时，进给运动是刀具的连续平移，而在刨削时，进给运动是刀具的间歇平移。图 1-15 所示为车削的基本加工参数。

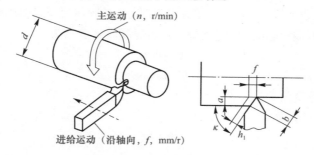

图 1-15　车削的基本加工参数

切削速度

切削速度 v 是刀具相对于工件的主要运动的瞬时速度（在切削刃上的选定点）。车削、钻削和铣削工序的切削速度可表示为

$$v = \pi dn \tag{1-1}$$

式中，v 为切割速度，单位为 m/min；d 为待切割工件的直径，单位为 m；n 为工件或主轴的转速，单位为 r/min。因此，根据具体的运动模式，v、d 和 n 可能与工件材料或与工具有关。在磨削中，切削速度通常用 m/s 来测量。

进给运动

当添加到主运动时，进给量 f 提供给刀具或工件，导致重复或连续的切屑去除并产生所需的加工表面。运动可以循序渐进，也可以连续进行。进给速度 v_f，定义为模具进给运动相对于工件（在切削刃上的选定点）的瞬时速度。

对于车削和钻孔，进给量 f 可以按工件或刀具毫米每转（mm/r）测量；对于刨削和成型，f 是按刀具或工件毫米每冲程（mm/冲程）来测量的。铣削时，进给量 f_z 按刀具毫米每齿（mm/齿）测量；即 f_z 为工件在连续两次齿形切削作用之间的位移。因此，工作台的进给速度 v_f(mm/min) 是刀具的齿数 z、刀具的每分钟转数 n 和每齿进给量 f_z 的乘积（$v_f = nzf_z$）。

包含主运动和进给运动方向的平面被定义为工作平面，因为它包含负责切割动作的运动。

切削深度（后啮合）

在车削（图 1-15）中，切削深度 a（也称为啮合）是切削刃啮合或凸出工件原表面以下的距离。切割的深度决定了工件的最终尺寸。在车削中，采用轴向进给时，切削深度是工件半径减小的直接尺寸，而采用径向进给时，切削深度等于工件长度的减小量。

切屑厚度

未变形状态下的切屑厚度 h_1 是垂直于切削刃和垂直于切削方向的平面上测得的切屑厚度（图 1-15）。切削后的切屑厚度（即实际切屑厚度 h_2）大于未变形的切屑厚度，这意味着切削比例或切屑厚度比例（$r = h_1/h_2$）总是小于 1。

切屑宽度

未变形状态下的切屑宽度 b 是沿切削刃在垂直于切削方向的平面上测得的。

切割面积

对于单点刀具加工，切割面积 A 是未变形切屑厚度 h_1 与切屑宽度 b 的乘积（即 $A=h_1 b$），切割面积也可以用进给量 f 和切削深度 a 表示为

$$h_1 = f\sin\kappa$$
$$b = a/\sin\kappa \qquad (1-2)$$

式中，κ 为刀具刃口角。

因此，切割面积为

$$A = fa \qquad (1-3)$$

Evaluate

任务名称		金属切削过程		姓名	组别	班级	学号	日期	
考核内容及评分标准					分值	自评	组评	师评	平均分
三维目标	知识	切削速度、进给运动、切割深度等概念			25分				
	技能	能阅读专业英语文章、翻译专用英语词汇			40分				
	素养	在学习过程中秉承科学精神、合作精神			35分				
加分项	收获（10分）	收获（借鉴、教训、改进等）：			你进步了吗？			加分	
					你帮助他人进步了吗？				
	问题（10分）	发现问题、分析问题、解决方法、创新之处等：						加分	
总结与反思								总分	

2.3 Milling Operations

Text

After lathes, milling machines are the most widely used for manufacturing applications. In milling, the workpiece is fed into a rotating milling cutter, which is a multi-edge tool, as shown in Fig. 1-16. Metal removal is achieved through combining the rotary motion of the milling cutter and linear motion of the workpiece simultaneously.

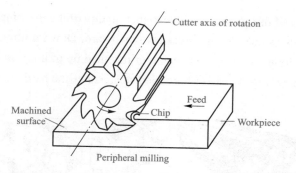

Fig. 1-16 Schematic diagram of a milling operation

Each of the cutting edges of a milling cutter acts as an individual single-point cutter when it engages with the workpiece metal. Therefore, each of those cutting edges has appropriate rake and relief angles. Since only a few of the cutting edges are engaged with the workpiece at a time, heavy cuts can be taken without adversely affecting the tool life. In fact, the permissible cutting speeds and feeds for milling are three to four times higher than those for turning or drilling. Moreover, the quality of surfaces machined by milling is generally superior to the quality of surfaces machined by turning, shaping, or drilling.

As far as the directions of cutter rotation and workpiece feed are concerned, milling is performed by either of the following two methods.

Up Milling (Conventional Milling) In up milling, the cutting tool rotates in the opposite direction to the table movement, the chip starts as zero thickness and gradually increases to the maximum size, as shown in Fig. 1-17(a). This tends to lift the workpiece from the table. There is a possibility that the cutting tool will rub the workpiece before starting the removal. However, the machining process involves no impact loading, thus ensuring smooth operation of the machine tool.

The initial rubbing of the cutting edge during the start of the cut in up milling tends to dull the cutting edge and consequently have a lower tool life. Also since the cutter tends to cut and slide alternatively, the quality of machined surface obtained by this method is not very high. Nevertheless, up milling is commonly used in industry, especially for rough cuts.

Down Milling (Climb Milling) In down milling, the cutting tool rotates in the same direction as that of the table movement, and the chip starts as maximum thickness and goes to zero thickness gradually, as shown in Fig. 1-17(b).

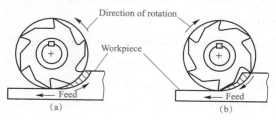

Fig. 1-17 Milling methods
(a) Up milling; (b) Down milling

The advantages of this method include higher quality of the machined surface and easier clamping workpieces, since cutting forces act downward. Down milling also allows greater feeds per tooth and longer tool life between regrinds than up milling. But, it cannot be used for machining castings (Fig. 1-18) or hot rolled steel, since the hard outer scale will damage the culler.

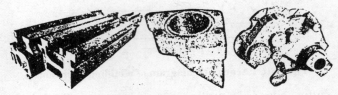

Fig. 1-18 Castings

There are a large variety of milling cutters to suit specific requirements. The cutters most generally used, shown in Fig. 1-19, are classified according to their general shape or the type of work they will do.

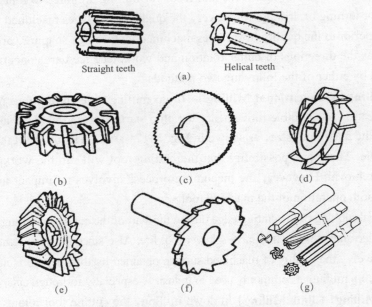

Fig. 1-19 Types of milling cutters

(a) Plain milling cutter; (b) Face milling cutter with inserted teeth; (c) Slitting saw; (d) Side and face cutter; (e) Angle milling cutter; (f) T□slot cutter; (g) End mill

Plain Milling Cutters

They are basically cylindrical with the cutting teeth on the periphery, and the teeth may be straight or helical, as shown in Fig. 1-19(a). The helical teeth generally are preferred over straight teeth because the tooth is partially engaged with the workpiece as it rotates. Consequently, the cutting force variation will be smaller, resulting in a smoother operation. The cutters are generally used for machining flat surfaces.

Face Milling Cutters

They have cutting edges on the face and periphery. The cutting teeth, such as carbide inserts, are mounted on the cutter body, as shown in Fig. 1-19(b).

Most larger-sized milling cutters are of inserted-tooth type. The cutter body is made of ordinary steel, with the teeth made of high speed steel, cemented carbide, or ceramics, fastened to the cutter body by various methods. Most commonly, the teeth are indexable carbide or ceramic inserts.

Slitting Saws

They are very similar to a saw blade in appearance as well as functions [Fig. 1-19(c)]. The thickness of these cutters is generally very small. The cutters are employed for cutting off operations and deep slots.

Side and Face Cutters

They have cutting edges not only on the periphery like the plain milling cutters, but also on both the sides [Fig. 1-19(d)]. As is the case with the plain milling cutter, the cutting teeth can be straight or helical.

Angle Milling Cutters

They are used in cutting dovetail grooves and the like. Fig. 1-19(e) indicates a milling cutter of this type.

T-slot Cutters

T-slot cutters [Fig. 1-19(f)] are used for milling T-slots such as those in the milling machine table.

End Mills

There are a large variety of end mills. One of the distinctions is based on the method of holding, the end mill shank can be straight or tapered. The straight shank is used on end mills of small size. The tapered shank is used for large cutter sizes. The culler usually rotates on an axis perpendicular to the workpiece surface.

Fig. 1-19(g) shows three kinds of end mills. The cutter can remove material on both its end and its cylindrical culling edges. Vertical milling machines and machining centers can be used for end milling of various sizes and shapes.

【 Words and phrases 】

milling cutter	铣刀
multi-edge tool	多刃刀具（也可以写为 multipoint tool）
schematic	*adj.* 示意性的；图解的；图表的
schematic diagram	原理图；示意图
peripheral milling	圆周铣削；周铣
cutting edge	切削刃；刀刃
single-point cutter	单刃刀具（切削时只用一个主切削刃的刀具）
rake angle	前角

relief angle	后角
tool life	刀具寿命
heavy cut	重切削；强力切削
up milling	逆铣
conventional milling	常规铣削
table	n. 工作台
table movement	工作台运动
machined surface	已加工表面
down milling	顺铣
climb milling	顺铣
chip	n. 切屑；片
regrind	v. 重新研磨；再次磨削
casting	n. 铸件；铸造
hot rolled steel	热轧钢
outer scale	外层氧化皮
plain milling cutter	普通铣刀
face milling cutter with inserted teeth	镶齿端铣刀
T-slot cutter	T形槽铣刀
end mill	立铣刀
helical teeth	螺旋齿
face milling cutter	端铣刀；面铣刀
carbide insert	硬质合金刀片
cutter body	刀体
cemented carbide	硬质合金
fasten	v. 固定；使坚固
indexable ceramic insert	可转位陶瓷刀片
slitting saw	锯片铣刀；切割锯
saw blade	锯条
side and face cutter	侧平两用铣刀
as is the case with	正如
angle milling cutter	角铣刀
dovetail groove	燕尾槽
and the like	等等
tapered shank	锥柄
milling machine	n. 铣床
machining center	加工中心

Course practice

1. Complex sentence analysis.

（1）In milling, the workpiece is fed into a rotating milling cutter, which is a multi-edge tool, as shown in Fig. 1-16.

be fed into：被输入，指物体、数据或信息被输入某个系统、设备或程序中。

【译文】在铣削过程中，工件被送入旋转的铣刀，这是一种多刃刀具，如图1-16所示。

（2）In fact, the permissible cutting speeds and feeds for milling are three to four times higher than those for turning or drilling.

be n times higher than…：……的 n 倍以上。

【译文】事实上，铣削允许的切削速度和进给量比车削或钻孔高 3～4 倍。

（3）In up milling, the cutting tool rotates in the opposite direction to the table movement…

in the opposite direction to：在相反的方向。

【译文】逆铣时，刀具旋转方向与工作台运动方向相反……

（4）They are basically cylindrical with the cutting teeth on the periphery, and the teeth may be straight or helical…

① be basically cylindrical：基本是圆柱形的。

② straight or helical：直的或是螺旋的。

【译文】普通铣刀基本呈圆柱形，切削齿在外围，切削齿可以是直的，也可以是螺旋的……

2. Translate the following sentences.

（1）The initial rubbing of the cutting edge during the start of the cut in up milling tends to dull the cutting edge and consequently have a lower tool life.

（2）Also since the cutter tends to cut and slide alternatively, the quality of machined surface obtained by this method is not very high.

3. Choose the proper answer to fill in the blank and translate the sentences.

(as, to, with, like, on, at, in)

（1）Each of the cutting edges of a milling cutter acts（　　）an individual single-point cutter.

（2）Moreover, the quality of surfaces machined by milling is generally superior（　　）the quality of surfaces machined by turning, shaping, or drilling.

（3）The helical teeth generally are preferred over straight teeth because the tooth is partially engaged（　　）the workpiece as it rotates.

Translation of the text

铣削加工

在车床之后,铣床是制造中应用最广泛的设备。在铣削过程中,工件被送入旋转的铣刀,这是一种多刃刀具,如图1-16所示。金属去除是通过同时结合铣刀的旋转运动和工件的直线运动来实现的。

图1-16 铣削操作示意

当铣刀与工件金属啮合时,铣刀的每个切削刃都充当单个单点铣刀。因此,每一个切削刃都有适当的前角和后角。由于一次只有少数切削刃与工件啮合,因此可以进行重切削,而不会对刀具寿命产生不利影响。事实上,铣削允许的切削速度和进给量比车削或钻孔高3~4倍。此外,铣削加工的表面质量通常优于车削、成型或钻孔加工的表面质量。

就刀具旋转方向和工件进给方向而言,铣削可通过以下两种方法中的一种进行。

逆铣(常规铣削) 逆铣时,刀具旋转方向与工作台运动方向相反,开始时切屑厚度为零,并逐渐增大到最大尺寸,如图1-17(a)所示。这使工件从工作台上被抬起。在开始去除(材料)之前,刀具有可能会摩擦工件。然而,加工过程中没有冲击载荷,从而保证了机床的平稳运行。

在逆铣切削开始时,切削刃的初始摩擦往往会使切削刃变钝,因此刀具寿命较短。此外,由于刀具倾向于交替切削和滑动,因此用这种方法获得的加工件表面质量不是很高。然而,逆铣在工业中还是常用的,特别是粗切。

顺铣 顺铣时,刀具旋转方向与工作台运动方向相同,切屑厚度从最大厚度开始,逐渐趋于零,如图1-17(b)所示。

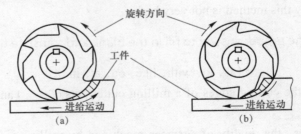

图1-17 铣削方法
(a) 逆铣;(b) 顺铣

这种方法的优点包括加工表面质量更高,且因为切削力向下作用,更容易夹紧工件。顺铣允许每齿更多的进给和更长的再次磨削之间的刀具寿命。但是,它不能用于加工铸件（图 1-18）或热轧钢，因为坚硬的外部水垢会损坏选线器。

图 1-18 铸件

有各种各样的铣刀来满足特定的要求。最常用的铣刀，如图 1-19 所示，是根据它们的一般形状或所执行的工作类型进行分类的。

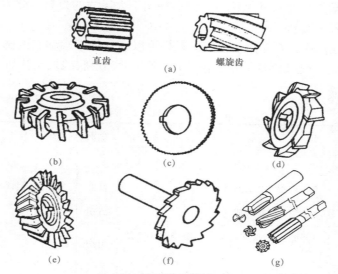

图 1-19 常用的铣刀

（a）普通铣刀；（b）镶齿面铣刀；（c）切割锯；（d）侧平两用铣刀；（e）角铣刀；（f）T形槽铣刀；（g）立铣刀

普通铣刀

普通铣刀基本呈圆柱形，切削齿在外围，切削齿可以是直的，也可以是螺旋的，如图 1-19（a）所示。螺旋齿通常优于直齿，因为部分螺旋齿可在旋转时与工件啮合。因此，切削力变化将更小，从而使操作更平稳。此刀具一般用于加工平面。

面铣刀

面铣刀的表面和周围都有锋利的边缘。如图 1-19（b）所示，切削齿（如硬质合金刀片）安装在刀体上。

大多数大型铣刀是嵌齿式的。刀体由普通钢制成，切削齿由高速钢、硬质合金或陶瓷制成，通过各种方法固定在刀体上。最常见的是，切削齿是可转位硬质合金或陶瓷刀片。

切割锯

切割锯在外观和功能上都与锯片非常相似［图 1-19（c）］。这些刀具的厚度通常很小，主要用于切断作业和切削深槽。

侧平两用铣刀

它们不仅像普通铣刀一样在外围有切削刃,而且在两侧都有切削刃[图1-19(d)]。与普通铣刀的情况一样,切削齿可以是直的,也可以是螺旋的。

角铣刀

角铣刀用于切割燕尾槽等。图1-19(e)所示为这种类型的铣刀。

T形槽铣刀

T形槽铣刀[图1-19(f)]用于铣削T形槽,如铣床工作台中的T形槽。

立铣刀

立铣刀种类繁多,其中一个区别是握住的方式不同,即立铣刀刀柄可以是直的或锥形的。直柄用于小尺寸立铣刀,锥形柄用于大尺寸的刀具。此刀具通常在垂直于工件表面的轴上旋转。

图1-19(g)显示了三种立铣刀,刀具可以在它的末端和它的圆柱形剔除边缘上去除材料。立式铣床和加工中心适用于各种尺寸和形状的末端铣削。

Evaluate

任务名称		铣削加工	姓名	组别	班级	学号	日期
考核内容及评分标准				分值	自评	组评	师评	平均分
三维目标	知识	逆铣、顺铣等概念		25分				
	技能	能阅读专业英语文章、翻译专用英语词汇		40分				
	素养	在学习过程中秉承科学精神、合作精神		35分				
加分项	收获(10分)	收获(借鉴、教训、改进等):	你进步了吗?		加分			
			你帮助他人进步了吗?					
	问题(10分)	发现问题、分析问题、解决方法、创新之处等:			加分			
总结与反思					总分			

2.4 Heat Treatment of Metal

Text

I. Heat Treatment Cycle Curve

The generally accepted definition for heat treating metals and metal alloys is "heating and cooling a solid metal or alloy in a way so as to obtain specific conditions and/or properties." Heating for the sole purpose of hot working (as in forging operations) is excluded from this definition. Likewise, the types of heat treatment that are sometimes used for products such as glass or plastics are also excluded from coverage by this definition.

Transformation Curves

The basis for heat treatment is the time-temperature-transformation curves or TTT curves where, in a single diagram, all the three parameters are plotted. Because of the shape of the curves, sometimes they are also called C-curves or S-curves.

To plot TTT curves, the particular steel is held at a given temperature and the structure is examined at predetermined intervals to record the amount of transformation taken place. It is known that the eutectoid steel (T8 steel) under equilibrium conditions contains all austenite above 727℃, whereas below, it is pearlite.

Classification of Heat Treating Processes

In some instances, the heat treatment procedures are clear in terms of technology and application. In other cases, however, description or simple explanation is not enough, as the same technique may often be used for different purposes. For example, T8 steel is carbon tool steel, which is a hardened plastic mold steel. Its chemical composition includes carbon (C), phosphorus (P), chromium (Cr), nickel (Ni), copper (Cu) and so on. T8 steel has high hardness and wear resistance after quenching and tempering, but low thermal hardness, poor hardenability, easy deformation, plasticity and strength. In order to improve the strength and toughness of T8 steel, a secondary heat treatment is usually required. By changing the temperature and duration of heat treatment to obtain the bainite we want. The heat treatment curve of T8 steel is shown in Fig. 1-20.

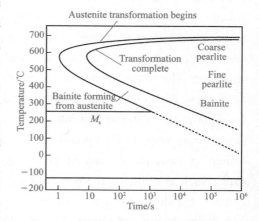

Fig. 1-20　Heat treatment curve of T8 steel

Of course, stress relieving and tempering can often also be done using the same equipment and using the same time and temperature cycles. However, the goals of the two processes are different.

II. Heat Treatment of Metals

Heat treatment is the operation of heating and cooling a metal in its solid state to change its physical properties. According to the procedure used, steel can be hardened to resist cutting action and abrasion, or it can be softened to permit machining. With the proper heat treatment internal stresses may be removed, grain size reduced, toughness increased, or a hard surface produced on a ductile interior.

The following discussion applies principally to the heat treatment of plain carbon steels. In this process, the rate of cooling is the controlling factor, and rapid cooling from above the critical range results in hard structure, whereas very slow cooling produces the opposite effect.

Annealing The primary purpose of annealing is to soften hard steel so that it may be machined or cold worked. This is usually accomplished by heating the steel to slightly above the critical temperature, holding it there until the temperature of the workpiece is uniform throughout, and then cooling at a slowly controlled rate so that the temperature of the surface and that of the center of the workpiece are approximately the same. This process is known as full annealing because it wipes out all trace of previous structure, refines the crystalline structure, and softens the metal. Annealing also relieves internal stresses previously set up in the metal.

Plain low carbon steels, when fully annealed, are soft and relatively weak, offering little resistance to cutting, but usually having sufficient ductility and toughness that a cut chip tends to pull and tear the surface from which it is removed, leaving a comparatively poor quality surface, which results in a poor machinability rating. For such steels annealing may not be the most suitable treatment. The machinability of many of the higher plain carbon steels and most of the alloy steels can usually be greatly improved by annealing, as they are often too hard and strong to be easily cut at any but their softest condition. The procedure for annealing hypoeutectoid steel is to heat it slowly to about 60 ℃ above the A_{c3} temperature, to soak for a long enough period that the temperature equalizes throughout the material and homogeneous austenite is formed, and then to allow the steel to cool very slowly by cooling it in the furnace or burying it in lime or some other insulating materials. For maximum softness and ductility the cooling rate should be very slow, such as allowing the parts to cool down with the furnace. The higher the carbon content, the slower this rate must be.

Normalizing and Spheroidizing The process of normalizing consists of heating the steel 10 ℃ to 40 ℃ above the upper critical range and cooling in still air to room temperature. This process is principally used with low and medium carbon steels as well as alloy steels to make the grain structure more uniform, to relieve internal stresses, or to achieve desired results in physical properties. Most commercial steels are normalized after being rolled or cast.

Spheroidizing is the process of producing a structure in which the cementite is in a spheroidal distribution. If a steel is heated slowly to a temperature just below the critical range and held there for a prolonged period of time, this structure will be obtained. The globular structure obtained gives improved machinability to the steel. This treatment is particularly useful for hypereutectoid steels that must be machined.

Hardening Most of the heat treatment hardening processes for steel is based on the production of high percentages of martensite. The first step, therefore, is that used for most of the other heat treating processes to produce austenite. Hypoeutectoid steels are heated to about 60 ℃ above the A_{c3} temperature and allowed to soak to obtain temperature uniformity and austenite homogeneity. Hypereutectoid steels are soaked at about 60 ℃ above the A_{c1} temperature, which leaves some iron carbide present in the material.

The second step involves cooling rapidly in an attempt to avoid pearlite transformation. The cooling rate is determined by the temperature and the ability of the quenching media to carry heat away from the surface of the material being quenched and by the conduction of heat through the material itself.

Heat Treating Quenching High temperature gradients contribute to high stresses that cause distortion and cracking, so the quench should only as extreme as is necessary to produce the desired structure. Care must be exercised in quenching that heat is removed uniformly to minimize thermal stresses. For example, a long slender bar should be end-quenched, that is, inserted into the quenching medium vertically so that the entire section is subjected to temperature change at one time. If a shape of this kind of bar were to be quenched in a way that caused one side to drop in temperature before the other, change of dimensions would likely cause high stresses producing plastic flow and permanent distortion.

Two special types of quenching are conducted to minimize quenching stresses and decrease the tendency for distortion and cracking. In both, the hardened steel is quenched in a salt bath held at a selected lower temperature before being allowed to cool. These processes, known as austempering and martempering, result in products having certain desirable physical properties.

Tempering Steel that has been hardened by rapid quenching is brittle and not suitable for most uses. By tempering or drawing, the hardness and brittleness may be reduced to the desired point for service conditions. As these properties are reduced there is also a decrease in tensile strength and an increase in the ductility and toughness of the steel. The operation consists of reheating quench-hardened steel to some temperature below the critical range followed by any rate of cooling. Although this process softens steel, it differs considerably from annealing in that the process lends itself to close control of the physical properties and in most cases does not soften the steel to the extent that annealing would. The final hardened steel obtained from tempering is called tempered martensite.

The magnitude of the structural changes and the change of properties caused by tempering depend upon the temperature to which the steel is reheated. The higher the temperature, the greater the effect, so the choice of temperature will generally depend on willingness to sacrifice hardness and strength to gain ductility and toughness. Reheating to a temperature below 100% has little noticeable effect on hardened plain carbon steel. For temperature between 100 ℃ and 200 ℃, there is evidence of some structural changes. For temperature above 200 ℃ marked changes in structure and properties appear. Prolonged heating at just under the A_{c1} temperature will result in a spheroidized structure similar to that produced by the spheroidizing process.

High-alloy air-hardened steels are never normalized, since to do so would cause them to harden and defeat the primary purpose.

Tool steels are normally spheroidized to improve machinability. This is accomplished by heating to a temperature between 1,380–1,400 ℉ (749–760 ℃) for carbon steels and higher for many alloy tool steels, holding the heat for 1–4 hours, and cooling slowly in the furnace.

Large dies are usually roughed out, then stress relieved and finish machined. This will minimize change of shape not only during machining but during subsequent heat treating as well. Welded sections will also have locked-in stresses owing to a combination of differential heating and cooling cycles as well as to changes in cross section. Such stresses will cause considerable movement in machining operations.

Tool steel is generally purchased in the annealed condition. Sometimes it is necessary to rework a tool that has been hardened, and the tool must then be annealed. For this type of anneal, the steel is heated slightly above its critical range and then cooled very slowly.

Case Hardening The addition of carbon to the surface of steel parts and the subsequent hardening operations are important phases in heat treating. The process may involve the use of molten sodium cyanide mixture, pack carburizing with activated solid material such as charcoal or coke, gas or oil carburizing.

【Words and phrases】

I

forge	v. 锻造
transformation	n. 变换；转换；相变
eutectoid	adj. 共析的
austenite	n. 奥氏体
pearlite	n. 珠光体
quenching	n. 淬火
stress relieving	消除应力；低温退火
tempering	n. 回火

II

anneal	v. 使（金属、玻璃等）退火（变坚硬）
	n. 退火；重组（DNA 链条）
full annealing	完全退火
crystalline	adj. 透明的；水晶般的；水晶制的
plain low carbon steel	普通低碳钢
machinability	n. 切削性；机械加工性
hypoeutectoid	adj. 亚共析的
homogeneous	adj. 同种类的；同性质的；齐性的
normalizing	n. ［冶］正火；正常化；［数］规格化
	v. 使正常化
spheroidizing	n. 球化处理
still	adj. 不动的；静止的
cementite	n. ［材］渗碳体；碳化铁
spheroidal	adj. 类似球体的；球状的
hypereutectoid	n. 过共析体；高共析质
hardening	n. 硬化；淬火
	v. 硬化；使强壮（harden 的 ing 形式）
martensite	n. ［材］马氏体；马登斯体；马丁散铁
homogeneity	n. 同质；同种；同次性
iron carbide	［无化］碳化铁；渗碳体
slender	adj. 苗条的；纤细的
permanent	adj. 永久的；永恒的
austempering	n. 等温淬火；奥式回火法
martempering	v. 马氏体回火法；分级淬火（martemper 的 ing 形式）
quench-hardened	adj. 淬火硬化的

Course practice

1. Complex sentence analysis.

（1）It is known that the eutectoid steel（T8）under equilibrium conditions contains all austenite above 727 ℃, whereas below, it is pearlite.

① that：引导主语从句。
② eutectoid steel：共析钢。
③ below：指低于 727 ℃。

【译文】已知在平衡条件下的共析钢（T8 钢）在 727 ℃以上时全部为奥氏体，而在 727 ℃以下为珠光体。

(2) Plain low carbon steels, when fully annealed, are soft and relatively weak, offering little resistance to cutting, but usually having sufficient ductility and toughness that a cut chip tends to pull and tear the surface from which it is removed, leaving a comparatively poor quality surface, which results in a poor machinability rating.

该句的基本句型为主语＋连系动词＋表语，句中 when fully annealed 作为时间状语，相当于在 when 后省略了 they are 的状语从句。

【译文】当完全退火时，普通低碳钢硬度降低、强度减弱，对切削的抵抗力很小，但通常由于塑性和韧性太大以致切屑离开工件表面时会划伤表面，工件表面质量较差，导致较差的切削加工性。

(3) The machinability of many of the higher plain carbon steels and most of the alloy steels can usually be greatly improved by annealing, as they are often too hard and strong to be easily cut at any but their softest condition.

① as...：该从句中有一个 too... to... 结构，译为太……而不能……
② but：除了。

【译文】许多高碳钢和大多数合金钢的切削加工性通常可通过退火大幅改善，除了在最软条件下之外，它们的硬度和强度太高而不易加工。

2. Translate the following paragraphs.

(1) Two special types of quenching are conducted to minimize quenching stresses and decrease the tendency for distortion and cracking. In both, the hardened steel is quenched in a salt bath held at a selected lower temperature before being allowed to cool.

(2) These processes, known as austempering and martempering, result in products having certain desirable physical properties.

3. Choose the proper answer to fill in the blank and translate the sentences.

(to, at, of, in, about, with)

(1) Heating and cooling a solid metal or alloy in a way so as (　　) obtain specific conditions and/or properties.

(2) The structure is examined at predetermined intervals to record the amount (　　) transformation taken place.

(3) In some instances, heat treatment procedures are clear cut in terms (　　) technique and application.

4. Answer the following questions.

(1) What is the purpose of full annealing?
(2) What is the word machinability used to describe?
(3) What is the purpose of normalizing?
(4) What does martempering mean?

(5) What does austempering mean?
(6) What is tempering?
(7) What process is used to remove the internal stresses created during a hardening operation?
(8) What heat treating process makes the metallic carbides in a metal form into small rounded globules?
(9) What are the main purposes of heat treating?
(10) How many heat treating processes are involved in ferrous materials?

5．Translate English words or phrases into Chinese according to the text.

(1) Plain low carbon steel　　　　　　(2) Hypoeutectoid steel
(3) Normalized steel　　　　　　　　(4) Hypereutectoid steel
(5) About 60 ℃ above the Ac1 temperature　(6) Cooling rate
(7) Quenching media　　　　　　　　(8) Thermal stress
(9) A tempered steel in the tempering or drawing procedure hardened steel full annealing

6．Translate the following words or phrases into English according to the text.

(1) 晶体的　　　　　　　　　　　　(2) 切削加工性
(3) 亚共析的　　　　　　　　　　　(4) 奥氏体
(5) 正火　　　　　　　　　　　　　(6) 渗碳体
(7) 球化（处理）　　　　　　　　　(8) 过共析的
(9) 马氏体　　　　　　　　　　　　(10) 珠光体
(11) 淬火　　　　　　　　　　　　　(12) 回火

Translation of the text

金属热处理

I. 热处理曲线

金属和金属合金热处理的公认定义是"以某种方式加热和冷却固体金属或合金，以获得特定的条件和/或性能"。以热加工为唯一目的的加热（如锻造操作）不包括在本定义中。同样，有时用于玻璃或塑料等产品的热处理类型也不包括在本定义的范围内。

转变曲线

热处理的基础是时间－温度－转变曲线或称 TTT 曲线，即在一个单一的图表中绘制了所有三个参数。由于曲线的形状不同，它们有时也称为 C 形曲线或 S 形曲线。

为了绘制 TTT 曲线，将特定的钢保持在给定的温度下，并以预定的时间间隔检查其结构，以记录发生的转变量。已知在平衡条件下的共析钢（T8 钢）在 727 ℃以上时全部为奥氏体,而在727 ℃以下为珠光体。

热处理工艺分类

在某些情况下,热处理程序在技术和应用方面是明确的。然而,在其他情况下,描述或简单地解释是不够的,因为同样的技术可能经常用来实现不同的目标。例如,T8钢属于碳素工具钢,是淬硬型塑料模具用钢。其化学成分包括碳(C)、磷(P)、铬(Cr)、镍(Ni)、铜(Cu)等。T8钢淬火回火后有较高的硬度和耐磨性,但热硬性低、淬透性差、易变形、塑性及强度较低。为提高T8钢的强度和韧性,通常需要进行二次热处理。通过改变热处理的温度和热处理持续时间,可获得想要的贝氏体。T8钢热处理曲线如图1-20所示。

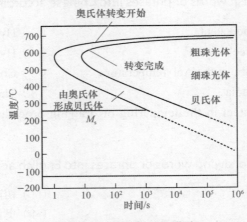

图1-20 T8钢热处理曲线

当然,应力消除和回火通常也可以使用相同的设备,并使用相同的时间和温度循环来完成。然而,这两个过程的目的是不同的。

II. 金属热处理

热处理是对固态金属进行加热和冷却以改变其物理特性的操作。根据使用的工艺,钢可以硬化以抵抗切削作用和磨损,也可以软化以便进行机械加工。通过适当的热处理,可以消除内应力,减小晶粒尺寸,增加韧性,或在易延展的内部形成坚硬的表面。

以下工艺主要适用于普通碳钢的热处理。在这个过程中,冷却速度是控制因素,从临界范围以上快速冷却会产生坚硬的结构,而缓慢冷却会产生相反的效果。

退火 退火的主要目的是软化硬钢,以便进行机械加工或冷加工。通常是通过将钢加热到略高于临界温度并保持,直到整个工件的温度均匀,然后缓慢控制冷却速度,使工件表面和中心的温度大致相同来实现的。这个过程被称为完全退火。因为它消除了以前结构的所有痕迹,细化了晶体结构,并软化了金属。退火还减小了先前在金属中形成的内应力。

当完全退火时,普通低碳钢硬度降低,强度减弱,对切削的抵抗力很小,但通常由于延展性和韧性太大以致切屑离开工件表面时会划伤表面,工件表面质量较差,导致较差的切削加工性。对于这种钢,退火可能不是最合适的处理方法。许多高碳钢和大多数合金钢的切削加工性通常可通过退火大幅改善，除了在最软条件下之外，它们的硬度和强度太高而不易加工。亚共析钢退火工艺是将其缓慢加热至 A_{c3} 温度以上约 60 ℃，保温足够长的时间，使整个材料的温度均衡，形成均匀的奥氏体，然后通过在炉中冷却，或将其埋在石灰中或一些其他绝缘材料中使钢非常缓慢地冷却。为了获得最大的柔软度和延展性，冷却速度应该非常缓慢，如让零件与炉子一起冷却。碳含量越高，这个速率就越低。

正火和球化 正火过程包括将钢加热至高于上临界范围 10～40 ℃，并在静止空气中冷却至室温。该工艺主要用于低碳钢和中碳钢以及合金钢，以使晶粒结构更加均匀，从而减小内应力，或在物理性能方面获得所需结果。大多数商品钢在轧制或铸造后都会进行正火处理。

球化是产生渗碳体呈球状分布结构的过程。如果将钢慢慢加热到刚好低于临界范围的温度，并保持很长一段时间，就会获得这种结构。所获得的球状结构改善了钢的可加工性。这种处理对于必须进行机械加工的过共析钢特别有用。

硬化 大多数钢的热处理硬化工艺都是基于高百分比马氏体的产生。因此，第一步是大多数其他热处理工艺所使用的，即生产奥氏体。将亚共析钢加热至高于 A_{c3} 温度约 60 ℃，并保温以获得温度均匀性和奥氏体一致性。过共析钢在高于 A_{c1} 温度约 60 ℃ 的温度下保温，从而在材料中留下一些碳化铁。

第二步是快速冷却，以免珠光体转变。冷却速率由温度和淬火介质从被淬火材料表面带走热量的能力，以及材料本身的热传导来决定。

热处理淬火 高温梯度会导致高应力，从而导致变形和开裂，因此淬火应仅在产生所需结构的特殊情况下进行。淬火时必须小心、均匀地去除热量，以最大限度地减小热应力。例如，细长棒材应进行端部淬火，即垂直插入淬火介质中，使整个截面一次性发生温度变化。如果这种形状的棒材被淬火，导致一侧温度先于另一侧温度下降，那么尺寸的变化可能会导致产生塑性流动和永久变形的高应力。

进行两种特殊类型的淬火，以最大限度地减小淬火应力并减少变形和开裂的趋势。在这两种情况下，硬化钢在冷却之前，保持在选定的较低温度的盐浴中淬火。这些工艺被称为奥氏体回火和马氏体回火，使产品具有某些理想的物理性能。

回火 通过快速淬火硬化的钢很脆，不适合大多数用途。通过回火或拉拔，硬度和脆性可以降低到使用条件所需的点。随着这些性能的降低，钢的抗拉强度也会降低，且延展性和韧性增加。该操作包括将淬火硬化钢重新加热到低于临界范围的某个温度，然后以任意速度冷却。尽管这种工艺可使钢软化，但它与退火有很大的不同，因为该工艺有助于对物理性能进行严格控制，并且在大多数情况下不会使钢软化到退火所能达到的程度。回火得到的最终硬化钢称为回火马氏体。

回火引起的结构变化和性能变化的幅度取决于钢被重新加热的温度。温度越高，影响越大，因此温度的选择通常取决于是否愿意牺牲硬度和强度以获得延展性和韧性。重新加热至低于 100% 的温度对硬化的普通碳钢几乎没有明显的影响；对于 100～200 ℃ 的温度，有证据表明存在一些结构变化；当温度超过 200 ℃ 时，会出现明显的结构和性能的变化。在 A_{c1} 温度下长时间加热将产生类似于球化过程产生的球化结构。

高合金空气硬化钢永远不会被正火，因为这样做会导致它们硬化并破坏其主要用途。

工具钢通常被球化处理以提高可加工性。这是通过将碳钢加热至 1 380～1 400 °F（749～760 ℃）的温度（对于许多合金工具钢，会加热至更高的温度），保温 1～4 h，并在炉中缓慢冷却来实现的。

大型模具通常先经过粗加工，然后进行应力消除和精加工。这不仅可以在加工过程中最大限度地减少形状的变化，而且在随后的热处理过程中，也能最大限度地减少形状的变化。由于不同的加热和冷却循环以及横截面的变化，焊接截面也将具有锁定应力。这种应力将在机械加工操作中引起相当大的移动。

工具钢通常在退火条件下供应。有时，有必要对已经硬化的工具进行返工，然后必须对该工具进行退火。对于这种类型的退火，钢被加热到略高于其临界范围的温度，然后非常缓慢地冷却。

表面硬化 向钢零件表面添加碳以及随后的硬化操作是热处理的重要阶段。该过程可能包括使用熔融的氰化钠混合物，用活性固体材料（如木炭或焦炭）进行填充渗碳、气体或油渗碳。

💬 Evaluate

任务名称		金属热处理	姓名	组别	班级	学号	日期
考核内容及评分标准			分值	自评	组评	师评	平均分
三维目标	知识	热处理曲线转化、退火、正火和球化等的概念	25 分				
	技能	能阅读专业英语文章、翻译专用英语词汇	40 分				
	素养	在学习过程中秉承科学精神、合作精神	35 分				
加分项	收获（10 分）	收获（借鉴、教训、改进等）：	你进步了吗？			加分	
			你帮助他人进步了吗？				
	问题（10 分）	发现问题、分析问题、解决方法、创新之处等：				加分	
总结与反思						总分	

Module 2

Electrical Control Technology

Focus

- Unit 1 Numerical Control Technology
- Unit 2 Electromechanical Integration Technology

Unit 1 Numerical Control Technology

Unit objectives

【Knowledge goals】
1. 掌握数控机床的原理、控制系统的类型等相关知识。
2. 掌握计算机数控（CNC）系统的概念、数控机床的优点和缺点等相关知识。
3. 掌握数控机床的坐标系统、数控机床的控制功能等相关知识。

【Skill goals】
1. 能够阅读专业英语文章。
2. 可以翻译专用英语词汇。

【Quality goals】
1. 在学习过程中发扬科学研究精神。
2. 增强团队合作意识。

1.1 Application of NC Technology

Text

One of the most fundamental concepts in the area of advanced manufacturing technologies is numerical control (NC).

Controlling a machine tool using a punched tape or stored program is known as numerical control. NC has been defined by the Electronic Industries Association (EIA) as "a system in which actions are controlled by the direct insertion of numerical data at some point. The system must automatically interpret at least some portion of this data". The numerical data required to produce a part is known as a part program.

A numerical control machine tool system contains a machine control unit (MCU) and the machine tool itself (Fig.2-1). The MCU is further divided into two elements, the data processing unit (DPU) and the control loops unit (CLU). The DPU processes the coded data from the tape or other media and passes information on the position of each axis, required direction of motion, feed rate, and auxiliary function control signals to the CLU. The CLU operates the drive mechanisms of the machine, receives feedback signals concerning the actual position and velocity of each of the axes, and signals of the completion of operation. The DPU sequentially reads

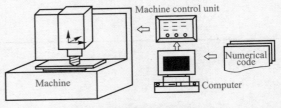

Fig. 2-1 Numerical control system

the data. When each line has completed execution as noted by the CLU, another line of data is read.

Geometric and kinematic data are typically fed from the DPU to the CLU and CLU then governs the physical system based on the data from the DPU.

Numerical control was developed to overcome the limitation of human operators, and it has done so. Numerical control machines are more accurate than manually operated machines. They can produce parts more uniformly, they are faster, and the long-run tooling costs are lower. The development of NC led to the development of several other innovations in manufacturing technology:

(1) Electric discharge machining;

(2) Laser-cutting;

(3) Electron beam welding.

Numerical control has also made machine tools more versatile than their manually operated predecessors. An NC machine tool can automatically produce a wide variety of parts, each involving an assortment of widely varied and complex machining processes. Numerical control has allowed manufacturers to undertake the production of products that would not have been feasible from an economic perspective using manually control led machine tools and processes.

I. Principles of NC Machines

An NC machine can be controlled through two types of circuits: open loop and close-loop. In the open loop system [Fig. 2-2(a)], the signals are sent to the servo motor by the controller, but the movements and final positions of the worktable are not checked for accuracy.

The close loop system [Fig. 2-2(b)] is equipped with various transducers, sensors, and counters that measure accurately the position of the worktable. Through feedback control, the position of the worktable is compared against the signal. Table movements terminate when the proper coordinates are reached. The close loop system is more complicated and more expensive than the open loop system.

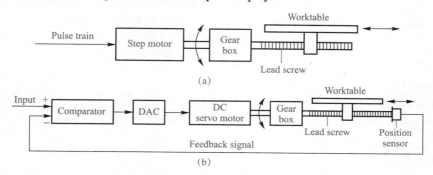

Fig. 2-2 Schematic illustration of NC machine

(a) An open loop control system for an NC machine; (b) A close loop control system for an NC machine

II. Types of Control Systems

There are two basic types of control systems in numerical control, point-to-point and contouring.

(1) In a point-to-point system (also called point positioning control system), each axis of the machine is driven separately by lead screws, and depending on the type of operation, at different velocities. The machine moves initially at maximum velocity in order to reduce nonproductive time, but decelerates as the tool approaches its numerically defined position. Thus, in an operation such as drilling (or punching a hole), the positioning and cutting take place sequentially.

After the hole is drilled or punched, the tool retracts upward and moves rapidly to another position, and the operation is repeated. The path followed from one position to another is important in only one respect. It must be chosen to minimize the time of travel, for better efficiency. Point-to-point systems are used mainly in drilling, punching, and straight milling operations.

(2) In a contouring system (also known as a continuous path system), the positioning and the operations are both performed along controlled paths but at different velocities. Because the tool acts as it travels along a prescribed path, accurate control and synchronization of velocities and movements are important. The contouring system is typically used on lathes, milling machines, grinders, welding machinery, and machining centers.

【Words and phrases】

numerical control (NC)	数字控制
advanced manufacturing technologies	先进制造技术
auxiliary	*adj.* 辅助的；备用的；后备的
	n. 助手；辅助人员
drive	*v. & n.* 驱动；驾驶
kinematic	*adj.* [力] 运动学上的；[力] 运动学的
tooling	*n.* 工具作业；压印图案；机床安装
	v. 使用工具；用工具加工
discharge	*v.* 释放；允许……离开；解雇；使退伍
	n. 允许出院；释放
versatile	*adj.* 多方面的；多变的；多用途的
assortment	*n.* 各种各样；混合
transducer	*n.* [自] 传感器；[电子] 变换器；[电子] 换能器
terminate	*v.* (使) 结束；(使) 终止；到达终点站
	adj. 结束的
close loop	闭环
open loop	开环
step motor	步进电动机

control system	控制系统
retract	vt. & vi. 缩回；缩进；取消
	n. 收缩核
synchronization	n. [物]同步；同时性
grinder	n. 磨床
machining center (MC)	加工中心

 Course practice

1. Complex sentence analysis.

（1）NC has been defined by the Electronic Industries Association (EIA) as "a system in which actions are controlled by the direct insertion of numerical data at some point. The system must automatically interpret at least some portion of this data".

①美国电子工业协会（EIA）：广泛代表了设计生产电子元件、部件、通信系统和设备的制造商以及工业界、政府和用户的利益，在提高美国制造商的竞争力方面起到了重要的作用。

②引号中的第一句为定语从句。

【译文】美国电子工业协会（EIA）对数字控制所下的定义："一种各项工作都由在各点上直接插入的数字来控制的系统。该系统必须至少能够自动解释这些数字中的部分数字。"

（2）The CLU operates the drive mechanisms of the machine, receives feedback signals concerning the actual position and velocity of each of the axes, and signals of the completion of operation.

① CLU：本句的主语，后面带三个动词作并列谓语。

② feedback signals：动词宾语。

③ concerning：引导的现在分词短语对其进行修饰。

④ signal：动词，表示发出信号，与前面两个动词时态相同，构成并列谓语。

【译文】CLU操作机床的机械驱动装置，接收关于每个轴的实际位置和速度反馈信号，并且发出操作完成信号。

（3）Numerical control has allowed manufacturers to undertake the production of products that would not have been feasible from an economic perspective using manually control led machine tools and processes.

① that：引导的定语从句对名词短语 production of products 进行修饰。

② from an economic perspective：从经济的观点出发，从经济的角度来看。

【译文】数字控制使制造者们可以进行产品的加工，从经济的角度来看其产品的加工使用人工控制机床和加工过程是不太可行的。

（4）The close loop system is more complicated and more expensive than the open-loop system.

more ... than ...：比……更……。

【译文】闭环系统比开环系统更复杂、更昂贵。

2. Translate the following paragraph.

Controlling a machine tool using a punched tape or stored program is known as numerical control. NC has been defined by the Electronic Industries Association (EIA) as " a system in which actions are controlled by the direct insertion of numerical data at some point. The system must automatically interpret at least some portion of this data ". The numerical data required to produce a part is known as a part program.

3. Choose the proper answer to fill in the blank and translate the sentences.

(into, from, on, to, with, through, against, of)

(1) The MCU is further divided (　　) two elements, the data processing unit (DPU) and the control loops unit (CLU).

(2) Geometric and kinematic data are typically fed (　　) the DPU to the CLU and CLU then governs the physical system based (　　) the data from the DPU.

(3) The signals are sent (　　) the servo motor by the controller, but the movements and final positions of the worktable are not checked for accuracy.

(4) The close loop system is equipped (　　) various transducers, sensors, and counters that measure accurately the position of the worktable.

 Translation of the text

数控技术

数字控制技术的应用

在先进制造技术领域，数字控制（NC）是最基本的理念之一。

使用冲孔纸带或存储程序控制机床称为数字控制。美国电子工业协会（EIA）对数字控制所下的定义："一种各项工作都由在各点上直接插入的数字来控制的系统。该系统必须至少能够自动解释这些数字中的部分数字。"用来加工一个零件的数据称为零件程序。

数控机床系统包含机床控制单元（MCU）和机床本身（图2-1）。机床控制单元进一步划分为两部分：数据处理单元（DPU）和控制回路单元（CLU）。DPU 处理来自纸带或其他介质的代码数据，并在每个轴的位置传递信息到 CLU，包括需要的运动方向、进给速率及辅助功能控制信号。CLU 操作机床的机械驱动装置，接收关于轴的实际位置和速度反馈信号，并且发出操作完成信号。DPU 按顺序读取数据，当 CLU 提示数据读取完一行后，则开始读取另一行数据。

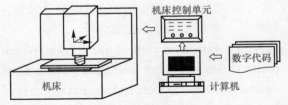

图2-1　数控机床系统

几何和运动数据一般从 DPU 传给 CLU，然后 CLU 根据来自 DPU 的数据控制物理系统。

发展数字控制用于克服人工操作的局限性，并且现在已经做到了这一点。数控机床比人工操作机床精度高一些。它们能够加工一致性更好的零件，加工速度更快，工艺成本更低。数字控制的发展带动了加工技术领域其他几种新技术的发展：

(1) 电火花加工；

(2) 激光切削；

(3) 电子束焊接。

数字控制也使数控机床比人工操作的机床功能更多。数控机床能自动加工各种零件，每种零件都涉及多种复杂的加工工艺。数字控制使制造者们可以进行产品的加工，从经济的角度来看其产品的加工使用人工控制机床和加工过程是不太可行的。

I. 数控机床的原理

数控机床通过两种基本回路控制：开环和闭环。在开环控制系统 [图 2-2 (a)] 中，信号通过控制器传送给伺服电动机，但对工作台的移动和最终位置并不进行精确检测。

闭环控制系统 [图 2-2 (b)] 带有各种精确检测工作台位置的传感器、检测元件及计数器。通过反馈控制，将工作台的位置与该信号相比较。当达到合适的位置坐标后，工作台停止移动。闭环系统比开环系统更复杂、更昂贵。

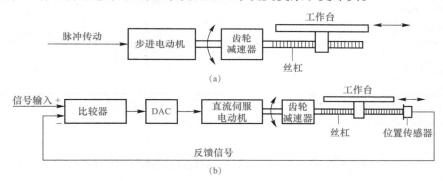

图 2-2 数控机床控制回路

(a)数控机床的开环控制系统；(b)数控机床的闭环控制系统

II. 数字控制系统的类型

数字控制有两种基本控制系统：点到点控制系统和轮廓控制系统。

(1) 在点到点控制系统（也称为点位控制系统）中，机床的每个轴由丝杠根据操作类型，以不同的速度单独驱动。为了减少非生产性时间，机床刚开始以最快的速度运动，但是当刀具接近它定义的位置时速度却减慢了。因此，在进行如钻孔（或冲孔）等操作时，定位和切削按照顺序执行。

孔被钻好或冲好后，刀具向上缩回并快速移动到另一位置，并重复操作。由一点到另一点的路径只有一项很重要，即为了更高效，必须选择移动时间最短的路径。点到点控制系统主要用于钻孔、冲孔和直线铣削加工。

(2) 在轮廓控制系统（也称为连续路径系统）中，以不同速度沿着控制路径进行定位和操作。由于刀具沿着指定路径运动，因此精确控制速度和位移的同步很重要。轮廓控制系统主要应用于车床、铣床、磨床、焊接机器和加工中心上。

Evaluate

任务名称	数控技术应用	姓名	组别	班级	学号	日期
考核内容及评分标准			分值	自评	组评	师评	平均分
三维目标	知识	数控机床的原理、控制系统的类型等相关知识	25 分				
	技能	能阅读专业英语文章、翻译专用英语词汇	40 分				
	素养	在学习过程中秉承科学精神、合作精神	35 分				
加分项	收获（10 分）	收获（借鉴、教训、改进等）：	你进步了吗？		加分		
			你帮助他人进步了吗？				
	问题（10 分）	发现问题、分析问题、解决方法、创新之处等：			加分		
总结与反思					总分		

1.2 Numerical Control Machines

Text

I. The Computer Numerical Control System

Numerical control is an automatic control method on the machine using digitized signals (numeric and signs). The NC concept was proposed in the late 1940s by John Parsons of Traverse City, Michigan. Numerical control is any machining process in which the operations are executed automatically in sequences as specified by the program that contains the information for the tool movements. In its earliest stages, NC machines were able to make straight cuts efficiently and effectively.

NC machines are known as machines with numerical control technology. According to the 5th Technical Committee of International Federation for Information Processing (IFIP), NC machine is a machine with procedures numerical control system, which can process the program prescript by specific code and other symbolic codes in a logic way.

Compared to the ordinary manual machine, the NC machine operates automatically under the program control (machining instructions). When numerical control is performed under computer supervision, it is called computer numerical control (CNC) (Fig. 2-3). CNC system has

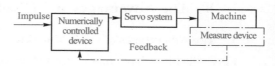

Fig. 2-3 Components of CNC

the computer system instead of a numerical control device. As defined by the Eletronic Industry Alliance (EIA) owned CNC Standardization Committee, CNC is a computer with a stored program, performing some or all of a the NC device features according to the control program stored in the computer read/write memory; the only device outside computer is the interface. CNC system is composed of procedures, input and output devices, computer numerical control devices, programmable logic controller (PLC), the spindle drive and feed drive device.

Computers are the control units of CNC machines, they are built in or linked to the machines via communication channels. When a programmer inputs some information in the program by tap disk and so on, the computer calculates all necessary data to get the job done.

Types of NC Machines

Since its introduction, NC technology has found many applications, including lathes and turning centers, milling machines and machining centers, punches, electrical discharge machines (EDM), flame cutters, grinders, and testing and inspection equipment. The most complex CNC machine is the turning center (Fig. 2-4), and the machining center (MC) (Fig. 2-5) (vertical-spindle machining center, with the tool magazine on the left and the control panel on the right, which can be swiveled by the operator) and horizontal-spindle machining center (equipped with an automatic tool changer) (Fig. 2-6). The EDM and flame cutter are special types of NC machines (Fig. 2-7 and Fig. 2-8).

Fig. 2-4 A modern turning center

Fig. 2-5 A vertical-spindle machining center

Fig. 2-6 A horizontal-spindle machining center

Fig. 2-7 An EDM machine

Fig. 2-8 A flame cutter

Ⅱ. The Advantages and Disadvantages of CNC Machines

1. The Advantages of CNC Machines

CNC machines have many advantages over conventional machines, some of them are as follows.

(1) There is a possibility of performing multiple operations on the same machine in one set-up.

(2) Because of the possibility of simultaneous multi-axis tool movement, special profile tools are not necessary to cut unusual part shapes.

(3) The scrap rate is significantly reduced because of the precision of the CNC machine tools and lesser operator impact.

(4) It is easy to incorporate part design changes when CAD/CAM systems are used.

(5) It is easier to perform quality assurance by a spot-check instead of checking all parts.

(6) Production is significantly increased.

2. The Disadvantages of CNC Machines

(1) They are quite expensive.

(2) They have to be programmed, set up, operated, and maintained by highly skilled personnel.

Obviously, CNC machines have more advantages than disadvantages. The companies that adopt CNC technology are increasing their competitive edge. With development of new technologies, the cost per CNC unit is being cut further and more companies can afford CNC equipment, which enables them to answer the increasingly strong requirements for production speed and quality that competitive markets demand. In the future, the broader use of CNC machines will be one of the best ways to enhance automation in manufacturing.

Ⅲ. The Composition of CNC Machines and Functions of Each Part

1. CNC Machining Process

(1) Carry on the machining process study; determine the machining plan, technological parameter and displacement data according to the pattern.

(2) Write the machining program to work part with the prescriptive code and format, or build the machining program file for parts with the automatic programming software, and perform the CAD/CAM work.

(3) Input or output the program. The manual-written program will be input on the CNC machine operation panel; the program generated by software will be transferred through serial communication interface (such as RS232c, etc.) of the computer to the numerical control unit of CNC machine tool.

(4) The process program in the CNC numerical control unit will operate test run and tool path simulation etc.

(5) Run the program, complete parts machining through the proper operation of CNC machines.

The basic principle of CNC machine tools is as below. First, write program based on components pattern combined with processing technology. Then input the program to numerical control device through the keyboard or other input device (such as perforated paper tape, floppy disks, etc.), the numerical control device decodes, stores and interpolates the program, give the command signal to the servo system of all the coordinates, in order to drive servo motor rotation, hold the relative position of the tool and workpiece and carry out their movement through the transmission, ensure their position accuracy through the position detection feedback. At the same time, initiate other necessary supporting actions through the PLC system, such as automatic transmission, the automatic opens and stops of the cooling lubricant, the automatic clamp and release of the workpiece, the automatic change of tools, etc. All of these operations, together with the feed movement, guarantee the automatic machining of parts.

2. Composition of an CNC Machine and Functions of Each Part

An CNC machine comprises of six parts: control media, NC system, servo system, feedback devices, auxiliary devices and the machine body, as shown in Fig. 2-9.

(1) Control media, also known as the information carrier, is the physical medium between humans and computers, reflecting all the information in the NC.

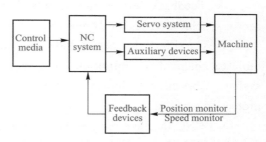

Fig. 2-9 Composition of the system of the CNC machine

(2) NC system is the core of the CNC machines, which ensure the automatic machining. It mainly consists of input devices, monitors, control system, programmable logic controller, input and output interfaces and so on.

Monitors are composed by the display and keyboard. Display could be digital LED, CRT, LCD and other forms. It shows the NC program, various parameters, interpolation value, coordinates, fault information, man-machine dialogue programming menu, part graphics, and dynamic cutting tool path.

The main control system consists of CPU, memory, controller and other components. There are two ways of control, computing process control and timing logic control.

Interpolation operation module in the master controller data can carry on the tool path interpolation operation after decoding and coding the input program, and control the displacement of the coordinate axis by comparison to the feedback signals of each axis' position and velocity.

The sequential logic control completes the task by programmable logic controller. It coordinates the requirements of each operation in the machining, determine the detection signals logically, and then control the machine tools to work methodically.

(3) Servo system is the electric transmission link between the numerical control system and machine body. It consists of a servo motor, drive control system, position sensing feedback device and other components. It is used to control the servo of feed and spindle.

① Feed servo system is the executive part of the feed movement, including the position control unit, speed control unit, servo motor, measuring feedback unit and other parts. It receives various commands of moves from computer to drive the servo motor movement. There are three types of servo motors, step motors, DC servo motors and AC servo motors. Performance of feed servo system has immediate effects on the machining accuracy and production efficiency of CNC machine tools.

② Spindle servo system is the part that transmits torque in cutting machine, generally divided into gear variable speed and electric continuously variable speed. Spindle servo pump is composed of the spindle drive control system, the spindle motor, and spindle mechanical transmission etc.

(4) Feedback devices include optical pulse encoder, grating position sensor, linear sensor synchronizer and other devices.

(5) Auxiliary devices include automatic tool changer, automatic change pallet table mechanical, damping and release mechanical, rotary table, hydraulic control system, lubrication equipment, cutting fluid devices, chip removal device, overload and protection devices.

(6) CNC machine body is the mechanical structure of ontology entities, composed by the main transmission, table, bed, spindle and other components. Compared with ordinary machines CNC machines have a sweeping change in its overall layout, appearance, transmission, tool system and operating mechanism, specifically summarized as follows:

① Has the great transmission power, high stiffness, good vibration and small thermal deformation with high performance main drive spindle parts.

② Has the short transmission chain, simple mechanism, high driving precision, such as the ball screw, synchronous gear belt etc., so as to ensure the transmission accuracy.

③ Has the perfect tool automatic exchange and management system (especially the CNC machining center).

④ Has the mechanism of automatic exchange workpiece, clamping and releasing workpiece in CNC machining centers.

⑤ The machine itself has a very high dynamic and static stiffness with some high precision and small friction coefficient components, such as plastic coated guide rails, linear roller guides rail and static guides rail etc.

⑥ Totally enclosed cover. As the CNC machine processes in an automatic way, for safety's sake, the machine's machining parts are totally enclosed with lagging and moving door.

Ⅳ. CNC Machining Object

1. The Advantages and Characteristics of CNC Machining
(1) Processability of complex-surfaced workpiece.
(2) High machining precision, good size consistency.
(3) High production efficiency.
(4) Low labor intensity.
(5) Obvious economic benefits.
(6) Accurate cost calculation and production schedule.
(7) The foundation for CAD/CAM technology and advanced manufacturing.

2. The Application Range of CNC Machines
(1) The parts produced in much variety and little batch; the parts in new product trial.
(2) The parts with complex geometry.
(3) The parts demanding multi-machining procedures.
(4) The parts demanding expensive process equipment when use general machining.
(5) High-precision demanding parts.
(6) The parts need repeated modifying in technical design.
(7) The expensive key parts not allowed scraped.
(8) The parts require the shortest production cycle.

The application range of CNC machines is shown in Fig. 2-10.

The relationship between machining batch of various machines and their cost is shown in Fig. 2-11.

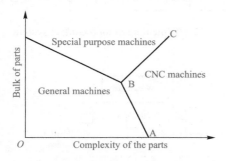

Fig. 2-10 The application range of various CNC machines

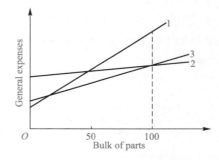

Fig. 2-11 The relationship between machining batch of various machines and their cost
1—General machine; 2—Special purpose machines
3—CNC machines

Ⅴ. The Construction of CNC Machines

CNC machines are complex assemblies. However, in general, any CNC machine consists of the following units:
(1) Computers;
(2) Control systems;

(3) Drive motors;
(4) Tool magazine and automatic tool changers.

1. Computers

The computer reacts on. As with all computers, the CNC machine computer works on a binary principle using only two characters 1 and 0, for information processing precise time impulses from the circuit. There are two states, a state with voltage, 1, and a state without voltage, 0. Series of 1s and 0s are the only states in which the computer distinguishes a so called machine language, it is the only language the computer understands. When creating the program, the programmer need not care about the machine language. He or she simply uses a list of codes and keys in the meaningful information. Special built-in software compiles the program into machine language and the machine moves the tool by its servo motors. However, the programmability of the machine is dependent on whether there is a computer in the machine's control. If there is a minicomputer programming, say, a radius (which is a rather simple task), the computer will calculate all the points on the tool path. On the machine without a minicomputer, this may prove to be a tedious task, since the programmer must calculate all the points of intersection on the tool path. Modern CNC machines use 32-bit processors in their computers that allow fast and accurate processing of information.

2. Control Systems

There are two types of control systems on NC/CNC machines, open loop (Fig. 2-12) and close loop (Fig. 2-13). The type of control loop used determines the overall accuracy of the machine.

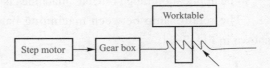

Fig. 2-12 Typical open loop control system

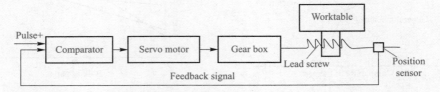

Fig. 2-13 Typical close loop control system

The open loop control system does not provide positioning feedback to the control unit. The movement pulses are sent out by the control unit and they are received by a special type of servo motor called a step motor. The number of pulses that the control unit sends to the step motor controls the amount of the rotation of the motor. The step motor then proceeds with the next movement command. Since this control system only counts pulses and cannot identify discrepancies in positioning, the machine will continue this inaccuracy until somebody finds the error.

The open loop control system can be used in applications in which there is no change in load conditions, such as the NC drilling machine. The advantage of the open loop control system is that it is less expensive, since it does not require the additional hardware and electronics needed for positioning feedback. The disadvantage is the difficulty of detecting a positioning error.

In the close loop control system, the electronic movement pulses are sent from the control unit to the servo motor, enabling the motor to rotate with each pulse. The movements are detected and counted by a feedback device called a transducer. With each step of movement, a transducer sends a signal back to the control unit, which compares the current position of the driven axis with the programmed position. When the numbers of pulses sent and received match, the control unit starts sending out pulses for the next movement.

Close loop systems are very accurate. Most have an automatic compensation for error, since the feedback device indicates the error and the control unit makes the necessary adjustments to bring the slide back to the position. They use AC, DC or hydraulic servo motors.

3. Drive Motors

The drive motors control the machine slide movement on NC/CNC equipment. They come in four basic types:

(1) Step motors;

(2) DC servo motors;

(3) AC servo motors;

(4) Hydraulic servo motors.

Step motors are used in open loop control systems, while AC, DC and hydraulic servo motors are used in close loop control systems.

4. Tool Magazine and Automatic Tool Changers

Most of the time, several different cutting tools are used to produce a part. The tools must be replaced quickly for the next machining operation. For this reason, the majority of NC/CNC machines are equipped with automatic tool changers, such as tool magazines on machining centers and turrets on turning centers.

They allow tool changing without the intervention of the operator. Typically, an automatic tool changer grips the tool in the spindle, pulls it out, and replaces it with another tool. On most machines with automatic tool changers, the turret or magazine can rotate in either direction, forward or reverse, shown in Fig. 2-14 and Fig. 2-15.

Automatic tool changers may be equipped for either random or sequential selection. In random tool selection, there is no specific pattern of tool selection. In the machining center, when the program calls for the tool, it is automatically indexed into waiting position, where it can be retrieved by the tool handing device. On the turning center, the turret automatically rotates, bringing the tool into position.

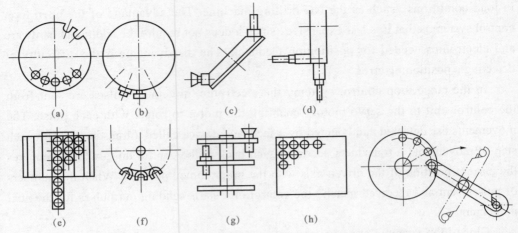

Fig. 2-14　A tool magazine

Fig. 2-15　An automatic tool changer

【Words and phrases】

I

computer numerical control（CNC）	计算机数字控制
numerical control（NC）	数字控制
impulse	n. 脉冲
servo system	伺服系统
measure device	测量装置
feedback	反馈
drive	v. & n. 驱动，驾驶
lathe	n. 车床
turning center	n. 车削加工中心
electrical discharge machine（EDM）	n. 电火花加工机床
flame cutter	n. 线切割机床
grinder	n. 磨床
vertical	adj. 立式的；垂直的
tool magazine	刀库
horizontal	adj. 水平的；卧式的
machining center	加工中心

II

advantage	n. 优点
conventional	adj. 传统的
set up	安装
perform	v. 执行

multiple		*adj.* 多样的
simultaneous		*adj.* 同时的
special profile tool		成型刀具
CAD（computer aided design）		计算机辅助设计
CAM（computer aided manufacturing）		计算机辅助制造
disadvantage		*n.* 缺点
maintain		*v.* 维护；保养

III

tool path		刀具路径
servo motor		伺服电动机
sensing		*n.* 传感
step motor		步进电动机
hydraulic		*adj.* 水力的；水压的；液压的 *n.* 液压传动装置
precision		*n.* 精度
stiffness		*n.* 刚性；僵硬；坚硬；不自然；顽固

IV

in general		一般地
consist		*v.* 包括；组成
automatic tool changer		自动换刀装置
binary		*adj.* 二进制的
minicomputer		*n.* 微型计算机
tedious		*adj.* 繁琐的
intersection		*n.* 交点
processor		*n.* 处理器
control system		控制系统
open loop		开环
close loop		闭环
discrepancy		*n.* 偏差
transducer		*n.* 传感器
hydraulic servo motor		液压伺服电动机；液压电动机
AC（alternative current）		交流
DC（direct current）		直流
turret		*n.* 转塔刀架
random selection		随机选刀
sequential selection		顺序选刀

Course practice

1. Complex sentence analysis.

（1）Numerical control is any machining process in which the operations are executed automatically in sequences as specified by the program that contains the information for the tool movements.

① in which：引导定语从句 the operations are executed automatically in sequences as specified by the program that contains the information for the tool movements，先行词是 machining process。

② that contains the information for the tool movements：作定语从句，修饰 the program。

【译文】数字控制是一种加工过程，其操作是按照包含刀具运动信息的程序指定的顺序自动执行的。

（2）Since its introduction, NC technology has found many applications ...

its：指代 NC technology。

【译文】自发明以来，数控技术得到了广泛的应用……

（3）CNC machines have many advantages over conventional machines ...

over：介词，译为超过，越过。

【译文】CNC 机床和传统机床相比具有许多优点……

（4）The scrap rate is significantly reduced because of the precision of the CNC machine and lesser operator impact.

because of：因为，由于……的原因。

【译文】由于 CNC 机床的精度高，受操作者的影响小，废品率明显降低。

（5）It is easy to incorporate part design changes when CAD/CAM systems are used.

it：作形式主语，真正的主语是 to incorporate part design changes。

【译文】当使用 CAD/CAM 系统时，很容易实现零件的设计变更。

（6）However, the programmability of the machine is dependent on whether there is a computer in the machine's control.

whether there is a computer in the machine's control：作介词 on 的宾语从句。

【译文】然而，机器的可编程性取决于其是否有计算机控制。

（7）The number of pulses that the control unit sends to the step motor controls the amount of the rotation of the motor.

that the control unit sends to the step motor：为定语从句，修饰 the number of pulses。

【译文】控制单元发送给步进电动机的脉冲数控制电动机的旋转量。

（8）They come in four basic types...

come in：归纳起来。

【译文】它们归纳起来有四种基本类型……

2. Translate the following paragraph.

(1) The open loop control system can be used in applications in which there is no change in load conditions, such as the NC drilling machine. The advantage of the open loop control system is that it is less expensive, since it does not require the additional hardware and electronics needed for positioning feedback. The disadvantage is the difficulty of detecting a positioning error.

3. Choose the proper answer to fill in the blank and translate the sentences.

(to, over, in, on, at, via, of, above)

(1) Compared (   ) the ordinary manual machine, the NC machine operates automatically under the program control.

(2) Computers are the control units of CNC machines, they are built (   ) or linked (   ) the machines via communication channels.

(3) CNC machines have many advantages (   ) conventional machines, some of them are as follows.

(4) Carry (   ) the machining process study; determine the machining plan, technological parameter and displacement data according (   ) the pattern.

Translation of the text

数控机床

I. 计算机数控系统

数字控制是一种利用数字化信号（数字和符号）对机器进行自动控制的方法。NC的概念是在20世纪40年代末由密歇根州特拉弗斯城的约翰·帕森斯提出的。数字控制是一种加工过程，其操作是按照包含刀具运动信息的程序指定的顺序自动执行的。在早期阶段，数控机床能够高效地进行直线切割。

数控机床被称为具有数控技术的机床。根据国际信息处理联盟（IFIP）第五技术委员会的定义，NC机床是一种具有程序数控系统的机床，它能以逻辑的方式加工特定代码和其他符号代码所规定的程序。

与普通手动机床相比，NC机床在程序控制（加工指令）下自动运行。当数控在计算机监督下进行时，就称为计算机数控（CNC）（图2-3）。CNC系统用计算机系统代替了数控装置。

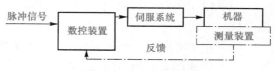

图2-3　CNC的组成

根据所属电子工业协会（EIA）的数控标准化委员会的定义，CNC是具有存储程序的计算机，根据存储在计算机读/写存储器中的控制程序，执行NC设备的部分或全部功能；计算机外部唯一的设备就是接口。CNC系统由程序、输入和输出装置、计算机数控装置、可编程逻辑控制器（PLC）、主轴驱动和进给驱动装置组成。

计算机是数控机床的控制单元，它们通过通信通道内置或与机床相连。当程序员通过敲击磁盘等方式在程序中输入一些信息时，计算机计算完成工作所需的所有数据。

数控机床的种类

自发明以来，数控技术已经得到了广泛的应用，包括车床和车削中心、铣床和加工中心、冲床、电火花加工机床（EDM）、火焰切割机、磨床，以及测试和检测设备。最复杂的 CNC 机床是车削中心（图 2-4）和加工中心（MC）（图 2-5）（立式主轴加工中心，左侧为刀库，右侧为控制面板，可由操作人员旋转）和卧式加工中心（配有自动换刀装置）（图 2-6）。EDM 和火焰切割机是特殊类型的 NC 机床（图 2-7 和图 2-8）。

图 2-4　现代车削中心

图 2-5　立式主轴加工中心

图 2-6　卧式主轴加工中心

图 2-7　电火花加工机床

图 2-8　火焰切割机

II. CNC 机床的优点和缺点

1. CNC 机床的优点

CNC 机床与传统机床相比具有许多优点，其中一些优点如下所示。

（1）可以在同一台机器上进行多种操作。

（2）由于可以同时进行多轴刀具移动，不需要特殊的轮廓刀具来切割特殊形状的零件。

（3）由于 CNC 机床的精度高，受操作者的影响小，废品率明显降低。

（4）当使用 CAD/CAM 系统时，很容易实现零件的设计变更。

（5）抽查比检查所有部件更容易保证质量。

（6）产量显著提高。

2. CNC 机床的缺点

（1）非常昂贵。

（2）必须由高技能人员编程、安装、操作和维护。

显然，CNC 机床的优点多于缺点。采用 CNC 技术的公司正在增强其竞争优势。随着新技术的发展，每台 CNC 设备的成本正在进一步降低，越来越多的公司能够负担得起 CNC 设备，这使他们能够满足竞争市场对生产速度和质量日益提升的要求。在未来，CNC 机床的广泛使用将是提高制造业自动化水平的最佳途径之一。

III. CNC 机床的组成及各部件的功能

1. 数控加工工艺

（1）进行加工工艺研究；根据图纸确定加工方案、工艺参数和排量数据。

（2）按照规定的代码和格式编写加工程序，或用自动编程软件编制零件加工程序文件，并执行 CAD/CAM 工作。

（3）输入或输出程序。手动编写的程序将输入到 CNC 机床操作面板上；软件生成的程序通过计算机的串行通信接口（如 RS232c 等）传输到 CNC 机床的数控单元。

（4）CNC 数控单元中的工艺程序将进行试车、刀具路径模拟等操作。

（5）运行程序，通过 CNC 机床的正确操作完成零件加工。

CNC 机床的基本原理如下。首先，结合加工工艺，编写基于组件的程序。然后通过键盘或其他输入设备（如穿孔纸带、软盘等）将程序输入数控设备，数控装置对程序进行解码、存储和插补，将所有坐标的指令信号传递给伺服系统，以驱动伺服电动机旋转，通过传动装置保持刀具和工件的相对位置并进行运动，通过位置检测反馈保证其位置精度。同时，通过 PLC 系统启动其他必要的配套动作，如自动传动、冷却润滑剂的自动开启和停止、工件的自动夹紧和释放、工具的自动更换等。所有这些操作配合进给运动，保证了零件的自动加工。

2. CNC 机床的组成及各部件的作用

CNC 机床由控制媒介、NC 系统、伺服系统、反馈装置、辅助装置和机床本体六部分组成，如图 2-9 所示。

（1）控制媒介，又称信息载体，是人与计算机之间的物理介质，反映 NC 中的一切信息。

（2）NC 系统是 CNC 机床的核心，保证了加工的自动化。它主要由输入设备、显示器、控制系统、可编程逻辑控制器、输入/输出接口等组成。

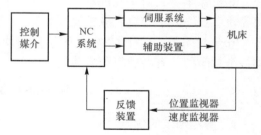

图 2-9 CNC 机床系统框图

显示器由显示屏和键盘组成，显示方式可采用数字 LED、CRT、LCD 等多种形式。它可以显示 NC 程序、各种参数、插补值、坐标、故障信息、人机对话编程菜单、零件图形、动态切削刀具轨迹等。

主控系统由 CPU、存储器、控制器等部分组成。控制方式有计算过程控制和定时逻辑控制两种。

主控制器数据中的插补操作模块对输入程序进行解码和编码后进行刀路径插补操作，并通过对比各轴位置和速度的反馈信号来控制坐标轴的位移。

顺序逻辑控制任务由可编程逻辑控制器完成。它协调加工中各个工序的要求，逻辑地确定检测信号，然后控制机床有条不紊地工作。

（3）伺服系统是数控系统与机床本体之间的电气传动环节。它由伺服电动机、驱动控制系统、位置感应反馈装置和其他部分组成。该系统用于控制进给和主轴的伺服。

①进给伺服系统是进料的执行部分，包括位置控制单元、速度控制单元、伺服电动机、测量反馈单元和其他部分。它接收来自计算机的各种动作指令，驱动伺服电动机运转。伺服电动机有三种类型，即步进电动机、DC 伺服电动机和 AC 伺服电动机。进给伺服系统的性能直接影响 CNC 机床的加工精度和生产效率。

②主轴伺服系统是切割机中传递扭矩的部分，一般分为齿轮变速和电动无级变速两种。主轴伺服泵由主轴驱动控制系统、主轴电动机、主轴机械传动等组成。

（4）反馈装置包括光脉冲编码器、光栅位置传感器、线性感应同步器等器件。

（5）辅助装置包括自动换刀装置、自动换盘机械装置、减振与释放机械装置、转盘、液压控制系统、润滑装置、切削液装置、排屑装置、过载及保护装置。

（6）机床本体是机械结构的本体实体，由主传动装置、工作台、床身和主轴等部件组成。与普通机床相比，CNC 机床在总体布局、外观、传动装置、刀具系统和操作机构等方面都发生了翻天覆地的变化，具体概括如下：

①传动功率大、刚度高、振动好、热变形小，并采用高性能主传动轴部件。

②传动链短、机构简单、传动精度高，并配有滚珠丝杠、同步齿轮带等良好的传动部件，保证传动精确度。

③具有完善的刀具自动交换和管理系统（特别是 CNC 加工中心）。

④在 CNC 加工中心一般具有自动交换工件、夹紧和释放工件的机构。

⑤机器本身具有很高的动、静刚度，有一些精度高、摩擦系数小的部件，如涂塑导轨、直线滚子导轨、静导轨等。

⑥全封闭盖。由于 CNC 机床是自动加工的，为了安全起见，机床的加工部件都是全封闭的，有滞后门和活动门。

Ⅳ．CNC 加工对象

1. 数控加工的优点

（1）复杂表面工件的可加工性。
（2）加工精度高、尺寸一致性好。
（3）生产效率高。
（4）劳动强度低。
（5）经济效益明显。
（6）准确的成本计算和生产计划。
（7）是 CAD/CAM 技术和先进制造的基础。

2. CNC 机床的应用范围

（1）产品品种多、批量小，新产品试制中的零件。

（2）几何形状复杂的零件。
（3）要求多道加工工序的零件。
（4）采用一般机械加工工艺时，需要昂贵的加工设备的零件。
（5）要求高精度的零件。
（6）在技术设计上需要反复修改的零件。
（7）昂贵的关键部件不允许刮擦。
（8）零件要求最短的生产周期。

各种 CNC 机床的应用范围如图 2-10 所示。

各种机床的加工批量与成本之间的关系如图 2-11 所示。

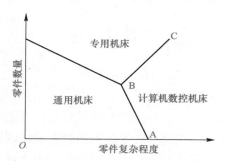

图 2-10　各种 CNC 机床的应用范围　　图 2-11　各种机床的加工批量与成本之间的关系

1—通用机床；2—专用机床；3—计算机数控机床

V．CNC 机床的构造

CNC 机床是复杂的装配体。但是，一般来说，任何 CNC 机床都由以下几个单元组成：

（1）计算机；
（2）控制系统；
（3）驱动电机；
（4）工具库和自动换刀器。

1．计算机

计算机会做出反应。与所有计算机一样，CNC 机床计算机使用二进制原理，只使用 1 和 0 两个字符，用于处理来自电路的精确时间脉冲的信息。字符有两种状态，有电压的状态为 1，无电压的状态为 0。一系列 1 和 0 是计算机区分所谓机器语言的唯一状态，这是计算机唯一能理解的语言。在编写程序时，程序员不必关心机器语言。他（或她）只是在有意义的信息中使用一系列代码和按键。特殊的内置软件将程序编译成机器语言，机器通过伺服电动机移动刀具。然而，机器的可编程性取决于其是否有计算机控制。如果由一个小型计算机编程，比如说，一个半径（这是一个相当简单的任务），计算机将计算刀具路径上的所有点。在没有微型计算机的机床上，这可能是一项繁琐的任务，因为程序员必须计算刀具路径上的所有交点。现代数控机床的计算机使用 32 位处理器，可以快速、准确地处理信息。

2. 控制系统

在 NC/CNC 机床上有两种类型的控制系统：开环（图 2-12）和闭环（图 2-13）。所使用的控制回路的类型决定了机床的整体精度。

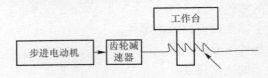

图 2-12　典型开环控制系统

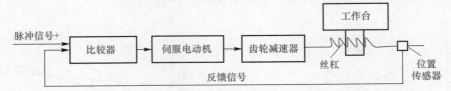

图 2-13　典型闭环控制系统

开环控制系统不向控制单元提供定位反馈。运动脉冲由控制单元发出，并由一种称为步进电动机的特殊类型的伺服电动机接收。控制单元发送给步进电动机的脉冲数控制电动机的旋转量。然后，步进电动机接着执行下一个运动命令。由于这个控制系统只计算脉冲数，不能识别定位中的差异，机器将继续这种不准确，直到有人发现错误。

开环控制系统可用于负载条件无变化的应用，如 NC 钻床。开环控制系统的优点是价格低，因为它不需要定位反馈所需的额外的硬件和电子设备，其缺点是难以检测定位误差。

在闭环控制系统中，电子运动脉冲从控制单元发送到伺服电动机，使电动机随着每个脉冲旋转。这些运动由一种称作换能器的反馈装置检测和计数。每移动一步，换能器向控制单元发送信号，控制单元将驱动轴的当前位置与编程位置进行比较。当发送和接收的脉冲数量匹配时，控制单元开始为下一个动作发送脉冲。

闭环系统非常精确。大多数系统都有一个自动补偿误差，因为反馈装置会显示误差，控制单元会做出必要的调整，使滑块回到正确的位置。它们使用 AC、DC 或液压伺服电动机。

3. 驱动电机

在 NC/CNC 设备上，驱动电机控制机床滑动。它们归纳起来有四种基本类型：

（1）步进电动机；

（2）DC 伺服电动机；

（3）AC 伺服电动机；

（4）液压伺服电动机。

步进电动机用于开环控制系统，而 AC、DC 和液压伺服电动机用于闭环控制系统。

4. 工具库和自动换刀器

大多数时候，生产一个零件要使用几种不同的刀具。为了下一个加工操作，必须迅速更换刀具。因此，大多数 NC/CNC 机床都配备了自动换刀装置，如加工中心上的工具库和车削中心的转台。

它们允许在没有操作人员干预的情况下更换刀具。通常情况下，自动换刀装置将刀具夹在主轴上，然后将其拔出，并用另一种刀具替换。在大多数带有自动换刀装置的机床上，转塔刀架或刀库可以在任何方向上旋转，向前或向后，如图2-14和图2-15所示。

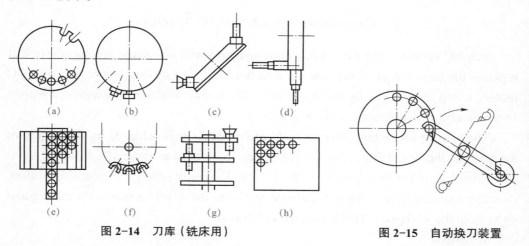

图2-14 刀库（铣床用）　　图2-15 自动换刀装置

自动换刀装置可用于随机选择或顺序选择。在随机工具选择中，没有特定的工具选择模式。在加工中心，当程序调用刀具时，刀具被自动分配到等待位置，在那里它可以被刀具处理装置取回。在车削中心，转塔刀架自动旋转，使刀具进入其位置。

Evaluate

任务名称		数控机床	姓名	组别	班级	学号	日期
		考核内容及评分标准	分值	自评	组评	师评	平均分
三维目标	知识	计算机数控（CNC）系统、数控机床的优点和缺点等相关知识	25分				
	技能	能阅读专业英语文章、翻译专用英语词汇	40分				
	素养	在学习过程中秉承科学精神、合作精神	35分				
加分项	收获（10分）	收获（借鉴、教训、改进等）：	你进步了吗？		加分		
			你帮助他人进步了吗？				
	问题（10分）	发现问题、分析问题、解决方法、创新之处等：			加分		
总结与反思					总分		

1.3 NC Programming

Text

I. Coordinate System for NC Machines

In a NC system, each axis of motion is equipped with a separate driving source that replaces the hand wheel of the conventional machine. The driving source can be a DC motor, a step motor, or a hydraulic actuator. The source selection is determined mainly on the precision requirements of the machine.

The relative movement between tools and workpieces is achieved by the motion of the machine slides. The three main axes of motion are referred to as the X, Y and Z axes. The Z axis is perpendicular to both the X and Y axes in order to create a right-hand coordinate system (Fig. 2-16). A positive motion in the direction moves the cutting tool away from the workpiece. This is detailed as follows.

1. Z-axis

(1) On a workpiece rotating machine, such as a lathe, the Z-axis is parallel to the spindle, and the positive motion moves the tool away from the workpiece (Fig. 2-17).

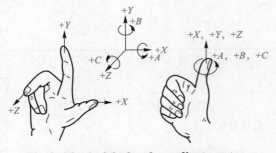

Fig. 2-16 A right-hand coordinate system

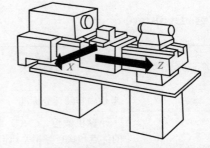

Fig. 2-17 Coordinate system for a NC lathe

(2) On a tool-rotating machine, such as a milling or boring machine, the Z-axis is perpendicular to the tool set, and the positive motion moves the tool away from the workpiece (Fig. 2-18 and Fig. 2-19).

(3) On other machines, such as a press machine, a planing machine, or a shearing machine, the Z-axis is perpendicular to the tool set, and the positive motion increases the distance between the tool and the workpiece.

2. X-axis

(1) On a lathe, the X-axis is the direction of tool movement, and positive motion moves the tool away from the workpiece.

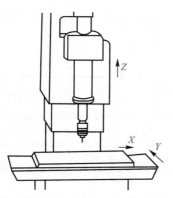
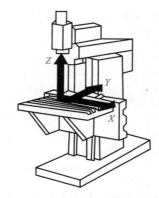

Fig. 2-18 A right-hand coordinate system   Fig. 2-19 Coordinate system for a NC lathe

(2) On a horizontal milling machine, the X-axis is parallel to the table.

(3) On a vertical milling machine, the positive X-axis points to the right when the programmer is facing the machine.

3. Y-axis

The Y-axis is the axis left in a standard Cartesian coordinate system.

II. The Control Functions of NC Machines

1. Control Functions of NC Machines

(1) Preparatory functions: the words specify which units, which interpolation, absolute or incremental programming, which circular interpolation plane, cutter compensation, and so on.

(2) Coordinates: coordinates define axes (including rotational).

(3) Machining parameters: machining parameters specify feed and speed.

(4) Tool control: tool control specifies tool diameter, next tool number, tool change, and so on.

(5) Cycle functions: cycle functions specify drilling cycle, reaming cycle, boring cycle, milling cycle, and clearance plane.

(6) Coolant control: coolant control specifies the coolant condition, that is, coolant on/off, flood, and mist.

(7) Miscellaneous control: miscellaneous control specifies all other control specifics, that is, spindle on/off, tape rewind, spindle rotation direction, pallet change, clamps control, and so on.

(8) Interpolators: include linear, circular interpolation, circle center, and so on.

2. Control Functions Are Programmed Through Program Words (Codes)

G-code   The G-code is also called preparatory code/word, it is used to prepare the microcontroller unit for control functions. Modal functions are those that do not change after they have been specified once, such as unit selection. Non-modal functions are active in the block where they are specified. For example, circular interpolation is a non-modal function. Some commonly used G-codes are listed in Table 2-1.

Table 2-1 G-code

Code	Meaning	Code	Meaning
G00	Rapid traverse	G40	Cutter compensation—cancel
G01	Linear interpolation	G41	Cutter compensation—left
G02	Circular interpolation (clockwise, CW)	G42	Cutter compensation—right
G03	Circular interpolation (counter clockwise, CCW)	G43	Cutter offset positive
G04	Dwell	G44	Cutter offset negative
G08	Acceleration	G80	Fixed cycle cancel
G09	Deceleration	G81-89	Fixed cycles
G17	$X-Y$ Plane	G90	Absolute dimension program
G18	$Z-X$ Plane	G91	Incremental dimension
G19	$Y-Z$ Plane	G92	Set the workpiece origin
G20	Inch format	G96	Constant surface speed control
G21	Metric format	G97	Constant spindle speed control
G28	Return to reference point	G98	Feed per minute
G33	Thread cutting	G99	Feed per revolution

F-code The F-code specifies the feed speed of the tool motion. It is the relative speed between the cutting tool and the workpiece. It is typically specified in mm/min, for example, F100 means the feed speed is 100 mm/min. Feed speed can be changed frequently in a program, as needed. When a F-code is present in a block, it takes effect immediately.

S-code The S-code is the cutting-speed code. Cutting speed is the specification of the relative surface speed of the cutting edge with relative to the workpiece. It is the result of the tool (or workpiece in turning) rotation. Therefore, it is programmed in rpm (r/min). The S-code is specified before the spindle is turned on. The S-code does not turn on the spindle. The spindle is turned on by an M-code. To specify an 800 r/min spindle speed, the NC program block is N0020 S800 M03.

T-code The T□code is used to specify the tool number. It is used only when an automatic tool changer is present. It specifies the slot number on the tool magazine in which the next tool is located. Actual tool change does not occur until a tool change M□code is specified.

R-code In a drill cycle, the R-code is used for cycle parameter. When a drill cycle is specified, one must give a clearance height (*R* plane) (Fig. 2-20). The R-code is used to specify this clearance height.

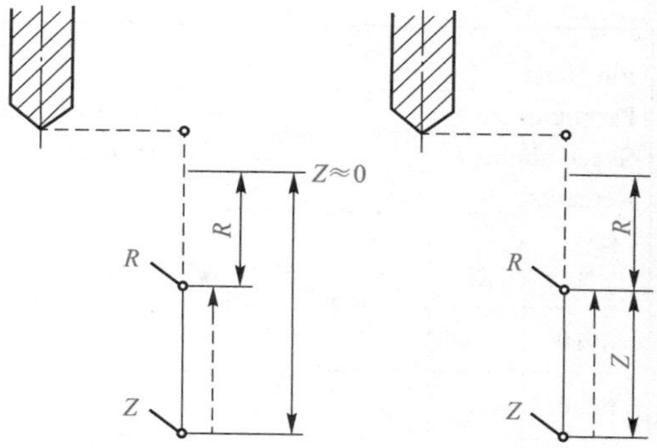

In Fig. 2-20, the drill cycle consists of five operations:

① Rapid location;

② Rapid down to the *R* plane;

③ Feed to the *Z* point, the bottom of the hole;

④ Operation at the bottom of the hole;

⑤ Rapid back or feed to either the *R* plane or the initial height.

Fig. 2-20 Drill cycle operation

The cycle may be programmed in one block:

N0010 G81 X1.000 Y2.000 Z0.000 R1.300

M-code The M□code is called the miscellaneous word and is used to control miscellaneous functions of the machine. Such functions (Table 2-2) include turning the spindle on/off starting/stopping the machine, turning on/off the coolant, changing the tool, and rewinding the program tape, etc.

Table 2-2 M-code

Code	Meaning	Code	Meaning
M00	Program stop	M08	Mist coolant on
M01	Optional stop	M09	Coolant off
M02	End of program	M10	Fixture close
M03	Spindle rotates (CW)	M11	Fixture open
M04	Spindle rotates (CCW)	M30	End of tape
M05	Spindle stop	M50	3# coolant on
M06	Tool changing	M51	4# coolant on
M07	Flood coolant on	M98	Calling of subprogram

III. NC Part Programming

In manual part programming, the machining instructions are recorded on a document, called a part-program manuscript (Table 2-3) by the part programmer. The manuscript is essentially an ordered list of program blocks.

Table 2-3   The NC part programming manuscript

Pin name _____
Part number _____
Sheet number _____
Remarks _____

Manuscript contouring program

Prepared by _____ Date _____
Checked by _____ Date _____
Machine number _____
Program number _____

N	G	X	Y	Z	I	J	K	F	S	T	M	Remark
0010												
0020												
0030												
0040												
0050												
0060												
0070												
0080												

Because a part program records a sequence of tool motions and operations to produce the final part geometry, one must prepare a process plan with setups and fixtures before writing the program. The workpiece location and orientation, features (holes, slots, pockets) to be machined, tools and cutting parameters used need to be determined. We will use an example to illustrate how a part is programmed.

See the example below.

The part drawing is shown in Fig. 2-21. The workpiece material is low carbon steel. We will use TH5632C machining center for the process. The process plan for the part is as follows.

(1) Set the center of the part as the machine zero point (floating-zero programming).

(2) Clamp the workpiece in a vise.

(3) Mill the profile with a 15 mm end mill made of carbide. From the *Machinability Date Handbook*, the recommended feed is 100 mm/min and the recommended cutting speed is 800 r/min. Fig. 2-22 shows the setup, and cutter path.

(4) During the milling operation, the cutter moves from A_1 to B_1, B_2, B_3, B_4, B_5, B_6, B_7, B_8 and then to B_1.

(5) The coordinates of each point (cutter location) are shown in Table 2-4:

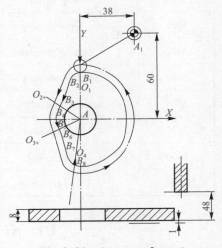

Fig. 2-21   An example part

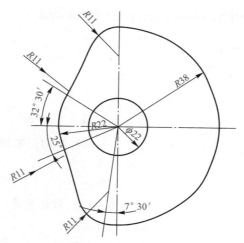

Fig. 2-22 The cutter path

Table 2-4 The coordinates of each point

O_1	(0, 27)	B_3	(-17.70, 13.46)
O_2	(-27.83, 17.73)	B_4	(-18.55, 11.82)
O_3	(-29.91, -13.95)	B_5	(-19.94, -9.30)
O_4	(-3.52, -26.77)	B_6	(-19.30, -10.99)
B_1	(0, 38)	B_7	(-14.12, -29.71)
B_2	(-10.12, 31.29)	B_8	(-4.96, -37.67)

Combining the information from the process plan and the cutter-location date, a part program can be written. The program for the example part is shown in Table 2-5.

Table 2-5 The program for the example part

N0010 G91 G17 G00 G41 H01 X - 38.0 Y - 42.0			S800 M03
N0020 G94	G01	Z - 38.0	F200
N0030	—	Z - 11.0	F100
N0040	G03 X - 10.12 Y - 6.7	—	R11.0
N0050	G01 X - 7.58 Y17.83	—	—
N0060	G02 X - 0.85 Y - 1.64	—	R11.0
N0070	G03 X - 1.39 Y21.12	—	R22.0
N0080	G02 X0.64 Y - 1.69	—	R11.0
N0090	G01 X5.18 Y - 18.71	—	—
N0100	G03 X9.16 Y7.96	—	R11.0
N0U0	X4.96 Y75.69	—	R-38.0
N0120	G00 G01	Z48.0	—
N0130	G40 X38.0 Y42.0	—	—
N0140	—	—	M30

We will use TH5632C machining center for the process.

【Words and phrases】

I

coordinate system	坐标系
step motor	步进电动机
actuator	n. [自] 执行机构；激励者；促动器；制动器
slide	v. & n. 滑行
perpendicular	adj. 垂直的
be referred to as...	被称为……；被认为是……
parallel	adj. 平行的
spindle	n. 主轴
positive	adj. 正方向的，正的
Cartesian coordinate system	笛卡儿坐标系

II

control function	控制功能
preparatory function	准备功能
interpolation	n. 插补
circular interpolation plane	圆弧插补平面
absolute	adj. 绝对的
incremental	adj. 增量的
cutter compensation	刀具补偿
parameter	n. 参数
cycle function	循环功能
fixed cycle	固定循环
coolant control	冷却控制
mist	v. 喷雾
miscellaneous control	杂项控制
pallet	n. 托盘
microcontroller unit（MCU）	微控制单元（MCU，又称单片机）
G-code	G 代码（准备功能指令）
M-code	M 代码（辅助功能指令）
S-code	S 代码（主轴转速指令）
T-code	T 代码（刀具指令）
F-code	F 代码（进给速度指令）
R-code	R 代码（循环功能指令）
clockwise（CW）	顺时针方向

counter clockwise（CCW）	逆时针方向
location	n. 定位

III

manual part programming	零件手工编程
manuscript	n. 单据，清单
block	n. 程序段
sequence	n. 次序
geometry	n. 几何学
illustrate	v. 说明
vise	n. 虎钳

Course practice

1. Complex sentence analysis.

（1）In a NC system, each axis of motion is equipped with a separate driving source that replaces the hand wheel of the conventional machine.

that replaces the hand wheel of the conventional machine：为定语从句，修饰 a separate driving source。

【译文】在 NC 系统中，每个运动轴都配备了一个单独的驱动源，以此来代替传统机床上的手轮。

（2）On a workpiece rotating machine, such as a lathe, the Z-axis is parallel to the spindle, and the positive motion moves the tool away from the workpiece...

a workpiece rotating machine：实际是指机床的主运动是由工件旋转而形成的。

【译文】在工件旋转的机床上，如车床，Z 轴是平行于主轴的，其正向运动使刀具远离工件……

（3）The words specify which units, which interpolator, absolute or incremental programming, which circular interpolation plane, cutter compensation, and so on.

which units, which interpolator, absolute or incremental programming, which circular interpolation plane, cutter compensation, and so on：为连续并列宾语。

【译文】指明程序段的单位、插补方式、绝对或增量编程、使用什么样的圆弧插补面、刀具补偿等。

（4）Modal functions are those that do not change after they have been specified once, such as unit selection.

that do not change after they have been specified once, such as unit selection：为定语从句，修饰先行词 those。

【译文】模态函数是指那些在指定一次之后不会改变的函数，如单位选择。

（5）The T-code is used to specify the tool number. It is used only when an automatic tool changer is present.

T代码即选刀指令，真正的换刀动作要等到M06指令出现时才发生。

【译文】T代码用于指定刀具编号，它只在有自动换刀装置时使用。

2．Translate the following sentences.

(1) The T-code is used to specify the tool number. It is used only when an automatic tool changer is present.

(2) It specifies the slot number on the tool magazine in which the next tool is located. Actual tool change does not occur until a tool change M-code is specified.

3．Choose the proper answer to fill in the blank and translate the sentences.

(to, between, at, of, about, in)

(1) The three main axes of motion are referred (　　) as the X-, Y-, and Z-axes.

(2) The Z-axis is perpendicular (　　) both the X- and Y-axes in order to create a right-hand coordinate system.

(3) On a horizontal milling machine, the X-axis is parallel (　　) the table.

(4) It is the relative speed (　　) the cutting tool and the workpiece.

(5) The M-code is called the miscellaneous word and is used (　　) control miscellaneous functions of the machine.

Translation of the text

数控编程

I．NC机床坐标系统

在NC系统中，每个运动轴都配备了一个单独的驱动源，以此来代替传统机床上的手轮。驱动源可以是DC电动机、步进电动机或液压执行器。驱动源选择主要根据机床的精度要求。

刀具和工件之间的相对运动是通过机床滑块的运动来实现的。三个主要的运动轴被称为X轴、Y轴和Z轴。Z轴垂直于X轴和Y轴，以创建一个右手坐标系（图2-16）。正向运动使刀具远离工件，具体如下。

1．Z轴

(1) 在工件旋转的机床上，如车床，Z轴与主轴平行，其正向运动使刀具远离工件（图2-17）。

(2) 在刀具旋转的机床上，如铣床或镗床，Z轴垂直于刀架，正向运动使刀具远离工件（图2-18和图2-19）。

(3) 在其他机床上，如压床、刨床或剪床，Z轴垂直于刀架，正向运动增加了刀具和工件之间的距离。

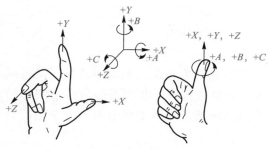
图 2-16 右手坐标系

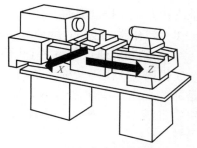
图 2-17 数控车削坐标系

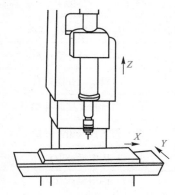
图 2-18 数控钻孔坐标系

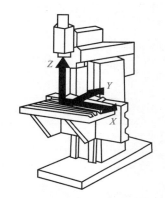
图 2-19 数控铣削坐标系

2. X 轴

（1）在车床上，X 轴是刀具运动的方向，正向运动使刀具远离工件。

（2）在卧式铣床上，X 轴与工作台平行。

（3）在立式铣床上，当程序员面向机床时，X 轴正方向指向右侧。

3. Y 轴

Y 轴是标准笛卡儿坐标系中的左轴。

Ⅱ. NC 机床的控制功能

1. NC 机床的控制功能

（1）准备功能：指明程序段的单位、插补方式、绝对或增量编程，使用什么样的圆弧插补面、刀具补偿等。

（2）坐标：坐标定义轴（包括旋转轴）。

（3）加工参数：加工参数规定进给和速度。

（4）刀具控制：刀具控制指定刀具直径、下一把刀具编号、换刀等。

（5）循环功能：循环功能指定钻孔循环、铰孔循环、镗孔循环、铣削循环和平面加工。

（6）冷却剂控制：冷却剂控制是指控制冷却剂的状态，即冷却剂的开/关、充满、雾化。

（7）杂项控制：杂项控制指定所有其他控制细节，即主轴开/关、胶带倒带、主轴旋转方向、托盘更换、夹具控制等。

（8）插补器：包括直线插补、圆弧插补、圆心插补等。

2. 控制功能通过程序字（代码）编程

G 代码 G 代码又称预备码 / 字，它用于准备具有控制功能的单片机。模态函数是那些在指定一次之后不会改变的函数，如单位选择。非模态函数在指定它们的块中是活动的。例如，圆插值是一个非模态函数。表 2-1 列出了一些常用的 G 代码。

表 2-1　G 代码

代码	含义	代码	含义
G00	快进	G40	刀具半径补偿：取消
G01	直线插补	G41	刀具半径补偿：左
G02	圆弧插补，顺时针	G42	刀具半径补偿：右
G03	圆弧插补，逆时针	G43	刀具长度正向补偿
G04	暂停	G44	刀具长度负向补偿
G08	加速	G80	固定循环取消
G09	减速	G81~89	固定循环开始
G17	X-Y 平面	G90	绝对尺寸程序
G18	Z-X 平面	G91	增量尺寸程序
G19	Y-Z 平面	G92	设置工件原点
G20	英寸格式	G96	恒线速控制
G21	米制格式（标准格式）	G97	主轴恒速控制
G28	返回参考点	G98	每分钟进给
G33	螺纹加工	G99	每转进给

F 代码 F 代码指定了刀具运动的进给速度，它是刀具与工件之间的相对速度，通常以 mm/min 为单位指定，如 F100 表示进给速度为 100 mm/min。根据需要，可以在程序中频繁更改进给速度。当一个 F 代码出现在一个区块中，它立即生效。

S 代码 S 代码是切削速度码。切削速度是指切削刃相对于工件的相对表面速度，它是刀具（或工件在车削时）旋转的结果。因此，它以 rpm（r/min）为单位编程。S 代码是在主轴开启前指定的。S 代码不启动主轴，主轴由 M 代码启动。要指定 800r/min 的主轴转速，NC 程序块为 N0020 S800 M03。

T 代码 T 代码用于指定刀具编号，它只在有自动换刀时使用。它指定了下一个刀具所在的刀具库上的槽号。在指定刀具更改 M 代码之前，不会发生实际的刀具更改。

R 代码 在钻孔循环中，R 代码用于循环参数。当指定钻孔循环时，必须给出间隙高度（R 平面）（图 2-20）。R 代码用于指定此间隙高度。

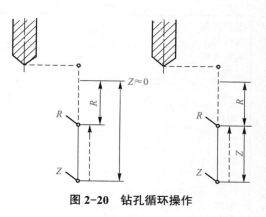

图 2-20 钻孔循环操作

在图 2-20 中，钻孔循环包括五项作业：

①快速定位；
②快速降至 R 平面；
③进给到孔底 Z 点；
④在孔底操作；
⑤快速返回或反馈到 R 面或初始高度。

这个循环可以在一个区块中编程：

N0010 G81 X1.000 Y2.000 Z0.000 R1.300

M 代码 M 代码称为杂项字，用于控制机器的杂项功能。这些功能（表 2-2）包括打开/关闭主轴、启动/停止机器、打开/关闭冷却液、换刀、倒带程序等。

表 2-2 M 代码

代码	含义	代码	含义
M00	程序停止	M08	雾化切削液开
M01	条件停止	M09	切削液关
M02	程序结束	M10	夹具打开
M03	主轴正转	M11	夹具关闭
M04	主轴反转	M30	主程序结束
M05	主轴停转	M50	3 号冷却液开
M06	换刀	M51	4 号冷却液开
M07	切削液开	M98	调用子程序

Ⅲ．NC 零件编程

在手工零件编程中，加工指令由零件编程人员记录在称为零件程序手稿（表 2-3）的文件上，手稿本质上是一个有序的程序块列表。

表 2-3　数控零件程序手稿

零件名称					轮廓程序			编制人：　　　　日期： 审核人：　　　　日期： 机床代号： 程序号：				
零件号												
表格代号												
备注												
N	G	X	Y	Z	I	J	K	F	S	T	M	标记
0010												
0020												
0030												
0040												
0050												
0060												
0070												
0080												

由于零件程序记录了一系列刀具运动和操作以产生最终零件的几何形状，因此必须在编写程序之前准备一个带有装置和夹具的工艺方案。需要确定工件的位置和方向、要加工的特征（孔、狭槽、刀槽）、使用的工具和切削参数。我们将用一个例子来说明零件是如何编程的。

举例如下。

零件如图 2-21 所示。工件材料为低碳钢。我们将使用 TH5632C 加工中心进行加工。该零件的工艺方案如下。

（1）设定零件的中心为机器零点（浮点零编程）。

（2）用虎钳夹住工件。

（3）用 15 mm 立铣刀铣削型材，刀具由硬质合金制成。根据《可加工性日期手册》建议进给量为 100 mm/min，建议切削速度为 800 r/min。图 2-22 显示了设置和刀具轨迹。

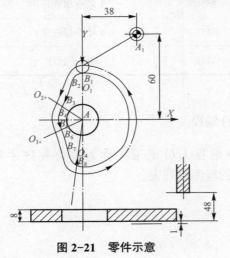

图 2-21　零件示意

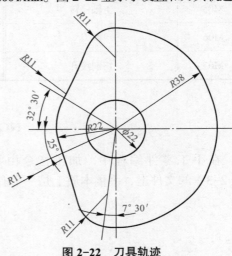

图 2-22　刀具轨迹

（4）铣削过程中，刀具从 A_1 移动到 B_1、B_2、B_3、B_4、B_5、B_6、B_7、B_8，再移动到 B_1。

（5）每个点坐标（刀具位置）见表 2-4。

表 2-4　每个点坐标

O_1	（0，27）	B_3	（-17.70，13.46）
O_2	（-27.83，17.73）	B_4	（-18.55，11.82）
O_3	（-29.91，-13.95）	B_5	（-19.94，-9.30）
O_4	（-3.52，-26.77）	B_6	（-19.30，-10.99）
B_1	（0，38）	B_7	（-14.12，-29.71）
B_2	（-10.12，31.29）	B_8	（-4.96，-37.67）

结合工艺计划和刀具定位数据的信息，可以编写零件程序。示例零件程序见表 2-5。

表 2-5　示例零件程序

N0010 G91 G17 G00 G41 H01 X-38.0 Y-42.0			S800 M03
N0020 G94	G01	Z-38.0	F200
N0030	—	Z-11.0	F100
N0040	G03 X-10.12 Y-6.7	—	R11.0
N0050	G01 X-7.58 Y17.83	—	—
N0060	G02 X-0.85 Y-1.64	—	R11.0
N0070	G03 X-1.39 Y21.12	—	R22.0
N0080	G02 X0.64 Y-1.69	—	R11.0
N0090	G01 X5.18 Y-18.71	—	—
N0100	G03 X9.16 Y7.96	—	R11.0
N0U0	X4.96 Y75.69	—	R-38.0
N0120	G00 G01	Z48.0	—
N0130	G40 X38.0 Y42.0	—	—
N0140	—	—	M30

我们将使用 TH5632C 加工中心进行加工。

Evaluate

任务名称		数控编程		姓名	组别	班级	学号	日期	
		考核内容及评分标准		分值	自评	组评	师评	平均分	
三维目标	知识	数控机床坐标系统、数控机床的控制功能等相关知识		25 分					
	技能	能阅读专业英语文章、翻译专用英语词汇		40 分					
	素养	在学习过程中秉承科学精神、合作精神		35 分					
加分项	收获（10 分）	收获（借鉴、教训、改进等）：	你进步了吗？				加分		
			你帮助他人进步了吗？						
	问题（10 分）	发现问题、分析问题、解决方法、创新之处等：					加分		
总结与反思							总分		

Unit 2　Electromechanical Integration Technology

Unit objectives

【Knowledge goals】
1. 掌握机电一体化的概念、在工业应用中的优缺点及其相关技术等知识。
2. 掌握单片机、中央处理器、堆栈指针等概念。
3. 掌握 PLC 的概念及其功能等知识。
4. 掌握传感器、无线传感技术、生物传感器等基本概念。
5. 掌握 CAD/CAM 相关的设计建模、数控编程、加工方法等知识。
6. 掌握与 CAE/CIM 相关的 CMOLD、级进模成型极限图（forming limit map, FLD）等知识。
7. 掌握工业机器人、笛卡儿坐标系、机器人的分类等知识。

【Skill goals】
1. 能够阅读专业英语文章。
2. 可以翻译专用英语词汇。

【Quality goals】
1. 在学习过程中发扬科学研究精神。
2. 增强团队合作意识。

2.1　Introduction to Mechatronics

机电一体化技术

Text

Japan coined the term of "Mechatronics" in 1970s to describe the integration of mechanical and electronic engineering. The concept is anything but new, since we can all look around us and see a myriad of products that utilize both mechanical and electronic disciplines. Mechatronics, specifically refers to a multidisciplinary, integrated approach to product and manufacturing system design. It represents the next generation of machines, robots and smart mechanisms necessary for carrying out work in a variety of environments—primarily, factory automation, office automation and home automation.

By both implication and application, mechatronics represents a new level of integration for advanced manufacturing technology and processes. The intent is to force a multidisciplinary approach to these systems as well as to reemphasize the role of process understanding and control. This mechatronics approach is currently speeding up the already rapid Japanese process for transforming ideas into products.

Currently, mechatronics describes the Japanese practice of using fully integrated teams of product designers, manufacturing, purchasing, and marketing personnel acting in concert with each other to design both the product and the manufacturing system.

The Japanese recognized that the future in production innovation would belong to those who learned how to optimize the marriage between mechanical and electronic systems. They realized, in particular, that the need for this optimization would be the most intense in application of advanced manufacturing and production systems where artificial intelligence, expert systems, smart robots, and advanced manufacturing technology systems would create the next generation of tools to be used in the factory of the future.

From the very beginning of recorded time, mechanical systems have found their way into every aspect of our society. Our simplest mechanisms, such as gears, pulleys, and wheels have provided the basis for our tools. Our electronics technology, on the other hand, completely belong to the twentieth century, all of which was created within the past 85 years.

Until now, electronics were included to enhance mechanical systems performance, but the emphasis remained on the mechanical product. There had never been any master plan on how the integration would be done. In the past, it had been done on a case-by-case basis. More recently, however, because of the overwhelming advances in the world of electronics and its capability to physically simplify mechanical configurations, the technical community began to reassess the marriage between these two disciplines.

The most obvious trend in the direction of mechatronics innovation can be observed in the automobile industry. There was a time when a car was primarily a mechanical marvel with a few electronic appendages.

Electronics has repeatedly improved the performance of mechanical systems, but that innovation has been more by serendipity than by design. That is the essence of mechatronics—the preplanned application and the efficient integration of mechanical and electronics technology to create an optimum product.

Comparisons were made in three categories: basic research, advanced development, and product implementation. In the advanced development and product implementation areas, Japan is equal to or better than the United States, and is continuing to pull ahead at this time.

To close the gap, we need to go much further than creating new tools. If we accept the fact that mechanical systems optimally coupled with electronic components will be the wave of the future, then we must also understand that the ripple effect will be felt all the way back to the university, where we now keep the two disciplines of mechanics and electronics separated and allow them to meet only in occasional overview sessions.

We need to rethink our present-day approach of separating our engineering staffs both from each other and from the production engineers. Living together and communicating individual knowledge will create a new synergistic effect on product development.

The definition of mechatronics is much more significant than its combined words. It can change the way we design and produce the next generation of high technology products. The nation that fully implements the rudiments of mechatronics and vigorously pursues then will lead the world to a new generation of technology innovation with all its profound implications.

Benefits of Mechatronics

Mechatronics may sound like Utopia to many product and manufacturing managers because it is often presented as the solution to nearly all of the problems in manufacturing. In particular, it promises to increase productivity in the factory dramatically. Design changes are easy with extensive use of mechatronics elements, such as CAD; CAP and MIS systems help in scheduling; and flexible manufacturing systems, computer aided design, and computer integrated manufacturing equipment cut turnaround time for manufacturing. These subsystems minimize production costs and greatly increase equipment utilization. Connections from CAE, CAD, and CAM help create designs that are economical to manufactures; control and communications are improved with minimal paper flow; and CAM equipment minimizes time loss due to setup and materials handling.

Many companies that make extensive use of computers view their factories as examples of mechatronics concepts, but on close examination their integration is horizontal in the manufacturing area only or at best includes primarily manufacturing and management. General Electric, as part of its effort to become a major vendor of factory automation systems, has embarked on ambitious plans for integration at several of its factories, including its Erie Locomotive Plant, its Schenectady Steam Turbine Plant and its Charlottesville Controls Manufacturing Division. The primary benefits of mechatronics, with an emphasis on advanced manufacturing technology and factory automation, are summarized below.

1. **High Equipment Utilization**

Typically, the throughput for a set of machines in a mechatronics system will be up to three times that for the same machines in a stand-alone job shop environment. The mechatronics system achieves high efficiency by having the computer schedule every part to a machine as soon as it is free, simultaneously moving the part on the automated material handling system and downloading the appropriate computer program to the machine. In addition, the part arrives at a machine already fixed on a pallet, this is done at a separate work station, so that the machine does not have to wait while the part is set up.

2. **Reduced Equipment Costs**

The high utilization of equipment results in the need for fewer machines in the mechatronic system to do the same work as in a conventional system. Reductions of 3 : 1 are common who replacing machining centers in a job shop situation with a mechatronics system.

3. **Reduced Direct Labor Costs**

Since each machine is completely under computer control, full-time oversight is not required. Direct labor can be reduced to the less skilled personnel who fix and release the parts at the station, and a machinist to oversee or repair the work stations, plus the system supervisor. While the fixing personnel in mechatronics environment require less advanced skills than corresponding workers, labor cost reduction is somewhat offset by the need for computing and other skills which may not be required in traditional workplaces.

4. Reduced Work-in-Process Inventory and Lead Time

The reduction of work-in-process in a mechatronics system is quite dramatic when compared to a job shop environment. Reductions of 80% have been reported at some installations and max be attributed to a variety of factors which reduce the time a part waits for metal-cutting operations. These factors include concentration of all the equipment required to produce part into a small area; reduction in the number of fixture required; reduction in the number of machines a part must travel through because processes are combined in work cells; and efficient computer scheduling of parts batched into and within the mechatronics system.

5. Responsiveness to Changing Production Requirements

A mechatronics system has the inherent flexibility to manufacture different products as the demands of the marketplace change or as engineering design changes are introduced. Further more, required spare part production can be mixed into regular runs without significantly disrupting the normal mechatronics system production activities.

6. Ability to Maintain Production

Many mechatronics systems are designed to degrade rationally when one or more machines fail. This is accomplished by incorporating redundant machining capability and a material handling system that allows failed machines to be bypassed. Thus, throughput is maintained at a reduced rate.

7. High Product Quality

Sometimes, an overlooked advantage of a mechatronics system especially when compared to machines that have not been federated into a cooperative system, is improved product quality. The basic integration of product design characteristics with production capability, the high level of automation, the reduction in the number of fixtures and the number of machines visited, better designed permanent fixtures, and greater attention to part/machine alignment all result in good individual part quality and excellent consistency from one workpiece to another, further resulting in greatly reduced cost of rework.

8. Operational Flexibility

Operational flexibility offers a significant increment of productivity. In some facilities, mechatronics systems can run virtually unattended during the second and third shift. This nearly unmanned, mode of operation is currently the exception rather than the rule. It should, however, become increasingly common as better sensors and computer controls are developed to detect and handle unanticipated problems such as tool breakages and part-flow jams. In this operational mode, inspection, fixture, and maintenance can be performed during the first shift.

9. Capacity Flexibility

With correct planning for available floor space, a mechatronics system can be designed for low production volumes initially; as demand increases new machines can be added easily to provide the extra capacity required.

【 Words and phrases 】

coin	v. 创造；杜撰（新词、新语）
term	n. 术语
mechatronics	n. 机械电子学；机电一体化
integration	n. 结合；一体化
discipline	n. 学科
integrate	v. 使一体化；使结合
generation	n.（一）代
smart	adj. 灵敏的；灵巧的
implication	n. 含义；本质
multidisciplinary	adj. 包括多种学科的
reemphasize	v. 反复强调
purchase	v. 购买；采购
market	v. 销售
innovation	n. 改革；革新
optimize	v. 使最佳化
optimization	n. 优化
performance	n. 性能；特性
overwhelming	adj. 压倒的；不可抵抗的
simplify	v. 简化
configuration	n. 结构；外形
community	n. 团体；界
reassess	v. 对……再评价
marriage	n. 结合；结婚
serendipity	n. 善于发掘新奇事物的才能
preplan	v. 预先计划；规划
implementation	n. 工具；仪器；执行过程
staff	n.（全体）工作人员
synergistic	adj. 叠加的；复合的
implement	v. 实现；提供方法
rudiment	n. 基础；入门
vigorously	adv. 强有力地
pursue	v. 推行；贯彻
profound	adj. 意义深远的
sound	v. 听起来
Utopia	n. 乌托邦；理想的完美境界

promise	v. 有……可能
dramatically	adv. 显著地
element	n. 组成部分
scheduling	n. 编制进度
cut	n. 削减
turnaround	n. 工作周期
subsystem	n. 子系统
minimize	v. 使……减至最小
utilization	n. 利用（率）
create	v. 产生；形成；创造
communication	n. 通信；联络
setup	v. 设置；安装
horizontal	adj. 横向的；水平的
vendor	n. 卖主
embark	v. 从事；着手
Erie	伊利（美国东部城市）
locomotive	n. 机车
Schenectady	斯克内克塔迪（美国东部城市）
turbine	n. 透平（机）；汽轮机；涡轮（机）
Charlottesville	夏洛茨维尔（美国南部城市）
division	n. 部门
summarize	n. 概括；摘要
throughput	n. 生产量
schedule	n. 安排；调度；排定
simultaneously	adv. 同时地
download	n. 下行传输；下载
fix	v. 夹紧；固定
	n. 夹具
pallet	n. 托盘
full-time	adv. 全部时间的；全部工作日的
oversight	n. 监督；观察
skilled	adj. 熟练的；需要技能的
machinist	n. 机械工人；机械师
oversee	n. 监视；照料
supervisor	n. 管理人员；检查员
skill	n. 技能；技术
offset	v. 抵消
workplace	n. 车间；工厂；工作面
in-process	adj. （加工、处理）过程中的

inventory	n.	库存；清单
installation	n.	装置
attribute	v.	由……引起
batch	n.	批；一批生产的量
responsiveness	n.	响应度；反应性
inherent	adj.	固有的；先天的
flexibility	n.	灵活性；适应性；柔性
introduce	v.	提出；引进；介绍
significantly	adv.	大大地；有意义地
disrupt	v.	中断；干扰
degrade	v.	降低（等级）；减少
fail	n.	失效；损坏
incorporate	v.	使结合
redundant	adj.	多余的
bypass	v.	绕过；走旁路
federate	v.	联合
cooperative	adj.	合作的；协作的
permanent	adj.	永久性的
alignment	n.	校正；调整
consistency	n.	一致性
rework	n. & v.	返工；再制
facility	n.	设备；设施
virtually	adv.	实质上；实际上
unattended	adj.	无人（看管）的
shift	n.	（轮班工作制的）班次
unmanned	adj.	无人（驾驶）的
mode	n.	方式
exception	n.	例外
increasingly	adv.	日益地
sensor	n.	传感器
handle	v.	处理
unanticipated	adj.	非预期的
breakage	n.	划伤
jam	n. & v.	阻塞；卡住
inspection	n.	检验；验收
maintenance	n.	维修；保养
floor space		工作空间
volume	n.	容量

Course practice

1. Complex sentence analysis.

（1）Japan coined the term of "Mechatronics" in 1970s to describe the integration of mechanical and electronic engineering.

the term of "Mechatronics"：" 机电一体化" 术语。

【译文】日本在20世纪70年代创造了术语"机电一体化"来描述机械和电子工程集成。

（2）It represents the next generation of machines, robots, and smart mechanisms necessary for carrying out work in a variety of environments—primarily factory automation, office automation, and home automation.

smart mechanism：智能机械/智能装置。

【译文】它代表了在各种环境中开展工作所需的下一代机器、机器人和智能机，主要是工厂自动化、办公自动化和家庭自动化的环境。

（3）This mechatronics approach is currently speeding up the already rapid Japanese process for transforming ideas into products.

speed up：加速。

【译文】目前，这种机电一体化方法正在加速日本快速将想法转化为产品的进程。

（4）Currently, mechatronics describes the Japanese practice of using fully integrated teams of product designers, manufacturing, purchasing, and marketing personnel acting in concert with each other to design both the product and the manufacturing system.

in concert with：协同，配合。

【译文】目前，机电一体化阐述了日本的实践方法，即通过产品设计师、制造人员、采购人员和营销人员组成的团队相互协作，设计产品和制造系统。

（5）The Japanese recognized that the future in production innovation would belong to those who learned how to optimize the marriage between mechanical and electronic systems.

optimize：v. 优化，充分利用（形势、机会、资源）；使（数据、软件等）优化；持乐观态度（optimise）。

【译文】日本人认识到，生产创新的未来将属于那些学会如何优化机械和电子系统之间关系的人。

（6）From the very beginning of recorded time, mechanical systems have found their way into every aspect of our society.

find way into：找到进入的方法，寻找进入某个地方或达到某个目标的方法或途径。

【译文】自从有记载以来，机械系统就进入我们社会的各个方面。

（7）There had never been any master plan on how the integration would be done.

　　integration：*n.* 结合，融合；取消种族隔离；（数）积分法，求积分。

【译文】对于如何集成，从未有过任何总体规划。

（8）The most obvious trend in the direction of mechatronics innovation can be observed in the automobile industry.

　　can be observed：可观察到。

【译文】在汽车工业中，可以观察到机电一体化创新方向最明显的趋势。

（9）Electronics has repeatedly improved the performance of mechanical systems, but that innovation has been more by serendipity than by design.

　　serendipity：*n.* 意外发现美好事物的运气，机缘巧合。

【译文】电子技术不断提高机械系统的性能，但这种创新更多的是出于偶然，而不是设计。

（10）Design changes are easy with extensive use of mechatronics elements, such as CAD; CAP and MIS systems help in scheduling; and flexible manufacturing systems, computer aided design, and computer integrated manufacturing equipment cut turnaround time for manufacturing.

① 此句是并列句，由三个简单句组成：第一句是 Design changes...such as CAD；第二句是 CAP and MIS system help in scheduling；第三句是 and flexible...for manufacturing。

② CAP（computer aided planning）：计算机辅助计划。

③ MIS（management information system）：管理信息系统。人员定期向此系统输入成本、生产、销售以及其他业务数据，并进行处理。有关问题分别提交上层管理部门进行决策。然后，有关信息又反馈到最高管理层，以反映实现重要目标的进度情况。

④ FMS（flexible manufacturing system）：柔性制造系统。该系统是利用计算机实现多级控制，以适应多品种、中小批量生产的一种高级自动化加工系统。柔性制造系统一般由主机（加工中心和其他数控机床）、连线设备（包括工业机器人、运输装置等）、控制设备（计算机等）及辅助设备组成。

⑤ CIM（computer integrated manufacturing）：计算机集成制造。计算机集成制造是指根据系统工程的观点，将整个车间或工厂作为一个系统，用计算机对产品的初始构思、设计、加工、装配和检验的全过程实行管理和控制。该系统的输入是产品需求的信息，输出为经过检验合格的产品。

【译文】广泛应用机电一体化的组成部分 CAD，使设计很容易被改变；CAP 和 MIS 系统可以帮助人们进行规划；柔性制造系统、计算机辅助设计和计算机集成制造等可以缩短整个制造过程的生产周期。

（11）Connections from CAE, CAD, and CAM help create designs that are economical to manufacture; control and communications are improved with minimal paper flow; and CAM equipment minimizes time loss due to setup and materials handling.

① 此句是并列句，由三个简单句组成：第一句是 Connections ... to manufacture；第二句是 control and ... paper flow；第三句是 and CAM ... materials handling。

② CAE（computer aided engineering）：计算机辅助工程。为检查一项设计的基本误差或使工艺性、经济性、生产率和性能达到最佳而使用信息处理系统的设计方法。

③ CAM（computer aided manufacturing）：计算机辅助制造。使用计算机来帮助制造业人员生产工业产品的加工体系。计算机用于制造业的两大方面：其一是面向工厂的管理和维护（包括仓库管理、计划调度、机器控制、信息管理和质量控制等），使管理与生产信息的处理连成一体；其二是面向制造工序的实际控制（包括对数控机床、其他加工机械、各种运输装置、机器人的控制），形成自动化工厂，使整个生产过程自始至终都处于计算机的控制之下。

【译文】将 CAE、CAD 和 CAM 结合起来，可以产生经济生产的设计；在最小的图纸流量下增强控制和交流；同时 CAM 设备可以减少安装和处理原材料的时间。

（12）The mechatronic system achieves high efficiency by having the computer schedule every part to a machine as soon as it is free, simultaneously moving the part on the automated material handling system and downloading the appropriate computer program to the machine.

① by having... simultaneously moving... and downloading... 是三个并列的动名词作介词 by 的宾语，与 by 组成短语，作方式状语。

② automatic material handling system：自动材料输送系统。

③ download：下行传输。把一个程序或数字文件从中央计算机向远程计算机或智能终端的存储器传输的过程。

【译文】当机器空闲下来时，就让计算机调度每个零件到这台机器，同时在自动材料输送系统中使零件运动，并且把适当的计算机程序下行传输到这台机器，这样就能使机电一体化系统达到很高的效率。

（13）In addition, the part arrives at a machine already fixed on a pallet, this is done at a separate work station, so that the machine does not have to wait while the part is set up.

① already fixed on a pallet：这是一个较长的定语，未跟在主语 the part 之后，以使句子结构匀称。

② work station：工作站（工位）。在制造系统中，零件正常停留的实际位置。在工作站，零件完成一道工序的加工或者在那里等待空位以便进入下一个工位。

【译文】此外，零件到达机器时，已经固定在托盘上，这是在一个单独的工作站完成的，这样机器就不必在零件安装时等待。

（14）Reductions of 3∶1 are common when replacing machining centers in a job-shop situation with a mechatronic system.

① machining center：加工中心，是一种采用计算机数控技术，集铣床、镗床、钻床三种功能于一体的加工机床，一般可进行铣、钻、镗、扩孔、铰孔、攻丝等工

序。可在一次装夹条件下，对工件的多面进行加工。加工中心配置有刀库和换刀装置。有的还配置了托板自动交换装置，以减少非切削时间，进一步提高生产率。加工中心有立式及卧式两种类型。

② replace … with … ：用……代替……。

【译文】在用机电一体化系统替换车间中的加工中心的情况下，成本通常是传统系统的 1/3。

（15）The reduction of work-in-process in a mechatronic system is quite dramatic when compared to a job shop environment.

work-in-process：在制工件，指处于加工过程中的零件，其中包括从刚进入加工过程的毛坯件到等待最终检验和入库的成品零件的整个生产过程中所有的加工工件。

【译文】与作业车间环境相比，机电一体化系统中在制品的减少是相当显著的。

（16）These factors include concentration of all the equipment required to produce part into a small area; reduction in the number of fixture required; reduction in the number of machines a part must travel through because processes are combined in work cells; and efficient computer scheduling of parts batched into and within the mechatronic system.

work cell：加工单元，由一个或多个工作站组成的制造设备。

【译文】这些因素包括：全部生产零件所需的设备都集中在一个小区域内；减少所需夹具数量；因为工序是在工作单元中组合的，所以应减少零件必须经过的机器数量；计算机有效调度零件批量进入机电一体化系统或在系统内调整。

（17）The basic integration of product design characteristics with production capability, the high level of automation, the reduction in the number of fixtures and the number of machines visited, better designed permanent fixtures, and greater attention to part/machine alignment all result in good individual part quality and excellent consistency from one workpiece to another, further resulting in greatly reduced cost of rework.

【译文】产品设计特点与生产能力的基本整合、高水平的自动化、夹具数量和机器访问数量的减少、设计更好的永久性夹具、更加关注零件与机器的调整，以及对零件/机器准线的更多关注都决定了良好的单个零件质量和从一个工件到另一个工件的卓越一致性，从而大幅降低了返工成本。

（18）Operational flexibility offers a significant increment of productivity. In some facilities, mechatronic systems can run virtually unattended during the second and third shift.

the second shift：（三班轮换制的）中班；the third shift：夜班；下文中有 the first shift，意思为早班（白班）。

【译文】操作的灵活性使生产效率明显提高。在一些使用机电一体化的工厂中，中班和晚班实际上是在无人看管的情况下运行的。

（19）With correct planning for available floor space, a mechatronics system can be designed for low production volumes initially; as demand increases new machines can be added easily to provide the extra capacity required.

① with：由于。
② plan for：作……计划。
③ floor space：占地面积。
④ production：产量。

【译文】通过对可用面积的正确规划，机电一体化系统在设计之初就可以满足小批量生产的需要；随着需求的增加，可以很容易地增加新机器以提供所需的额外产能。

2. Translate the following paragraph.

The most obvious trend in the direction of mechatronics innovation can be observed in the automobile industry. There was a time when a car was primarily a mechanical marvel with a few electronic appendages.

3. Choose the proper answer to fill in the blanks and translate the sentences.

（to, up, into, on, in, than, with, for, of, about, beyond）

（1）However, specifically refers（　　）a multidisciplinary, integrated approach to product and manufacturing system design.

（2）This mechatronics approach is currently speeding（　　）the already rapid Japanese process for transforming ideas（　　）products.

（3）Electronics were included（　　）enhance mechanical systems performance, but the emphasis remained（　　）the mechanical product.

（4）In the past, it had been done（　　）a case-by-case basis.

（5）Comparisons were made（　　）three categories: basic research, advanced development, and product implementation.

（6）The definition of mechatronics is much more significant（　　）its combined words.

（7）（　　）correct planning for available floor space, a mechatronics system can be designed（　　）low production volumes initially.

Translation of the text

机电一体化概论

日本在20世纪70年代创造了术语"机电一体化"来描述机械和电子工程集成。这个概念一点也不新鲜，因为我们可以看到周围无数同时利用机械和电子学科的产品。机电一体化，具体指的是用一种多学科的、综合的方法来进行产品和制造系统的设计。它代表了在各种环境中开展工作所需的下一代机器、机器人和智能机，主要是工厂自动化、办公自动化和家庭自动化的环境。

从内涵和应用上看，机电一体化代表了先进制造技术和工艺集成的新水平。其目的是促使对这些系统采用多学科方法，并重新强调过程理解和控制的作用。目前，这种机电一体化方法正在加速日本快速将想法转化为产品的进程。

目前，机电一体化阐述了日本的实践方法，即通过产品设计师、制造人员、采购人员和营销人员组成的团队相互协作，设计产品和制造系统。

日本人认识到，生产创新的未来将属于那些学会如何优化机械和电子系统之间关系的人。他们特别意识到，这种优化的需求将在先进制造和生产系统的应用中最为强烈，其中人工智能、专家系统、智能机器人和先进制造技术系统将创造下一代工具，用于未来的工厂。

自从有记载以来，机械系统就进入了我们社会的各个方面。我们最简单的构件，如齿轮、滑轮和轮子，为我们的工具提供了基础。另外，我们的电子技术完全属于20世纪，所有这些都是在过去85年里创造出来的。

到目前为止，电子产品用来提高机械系统的性能，但重点仍然是机械产品。对于如何集成，从未有过任何总体规划。在过去，总是根据具体情况进行整合。然而，最近，由于电子领域的巨大进步及其在物理上能简化机械结构的能力，技术界开始重新评估这两个学科之间的整合。

在汽车工业中，可以观察到机电一体化创新方向最明显的趋势。曾经有一段时间，汽车主要是带有电子附件的机械奇迹。

电子技术不断提高机械系统的性能，但这种创新更多的是出于偶然，而不是设计。这就是机电一体化的本质——机械和电子技术的预先计划应用及有效集成，以创造最佳产品。

从基础研究、先进开发和产品实施三个方面进行比较。在先进开发和产品实施领域，日本与美国相当或更胜一筹，并且在目前持续领先。

为了缩小差距，我们需要做的远不止创造新工具。我们如果接受这样一个事实，即机械系统与电子元件的最佳结合将是未来的潮流，那么也必须明白，这种连锁反应将一直波及大学，在大学，将力学和电子学两个学科分开，只允许在偶尔的概述会议上将其相提并论。

我们需要重新考虑目前的方法，即将工程技术人员与生产工程师彼此分开。共同生活和交流个人知识将为产品开发创造新的协同效应。

机电一体化的定义远比它的组合词更有意义，它可以改变我们设计和生产下一代高科技产品的方式。一个国家如果充分掌握机电一体化的基本原理，并大力推行机电一体化，那么它将引领世界走向新一代的技术创新，其意义深远。

机电一体化的好处

对于许多产品和制造经理来说，机电一体化可能听起来像乌托邦，因为它经常被认为是制造中几乎所有问题的解决方案。特别是，它有望大幅提高工厂的生产效率。广泛应用机电一体化的组成部分 CAD，使设计很容易被改变；CAP 和 MIS 系统可以帮助人们进行规划；柔性制造系统、计算机辅助设计和计算机辅助制造设备等可以缩短整个制造过程的生产周期。这些子系统最大限度地降低了生产成本，大幅提高了设备利用率。将 CAE、CAD 和 CAM 结合起来，可以产生经济生产的设计；在最小的图纸流量下增强控制和交流；同时 CAM 设备可以减少安装和处理原材料的时间。

许多广泛使用计算机的公司将他们的工厂视为机电一体化概念的典范，但仔细研究一下，他们的一体化只是横向的，仅涉及制造领域，或者充其量主要包括制造和管理。通用电气公司为了努力成为工厂自动化系统的主要供应商，已经开始实施计划，对其几个工厂进行整合，包括伊利机车厂、斯克内克塔迪汽轮机厂和夏洛茨维尔控制制造部门。机电一体化的主要优势是先进的制造技术和工厂自动化，将其总结如下。

1. 设备利用率高

通常，在机电一体化系统中，一组机器的产量将是独立作业车间中相同机器的三倍。当机器空闲下来时，就让计算机调度每个零件到这台机器，同时在自动材料输送系统中使零件运动，并且把适当的计算机程序下行传输到这台机器，这样就能使机电一体化系统达到很高的效率。此外，零件到达机器时，已经固定在托盘上，这是在一个单独的工作站完成的，这样机器就不必在零件安装时等待。

2. 降低设备成本

设备的高利用率决定了在机电一体化系统中需要较少的机器来完成与传统系统相同的工作。在用机电一体化系统替换车间中的加工中心的情况下，成本通常是传统系统的 1/3。

3. 降低直接人工成本

由于每台机器完全由计算机控制，因此不需要专职监督它们。直接人工标准可以降低为固定和释放零件的非技术人员，以及监督或修理工作站的机械师，再加上一名系统主管。虽然机电一体化环境下对固定人员的技能要求不高，但人工成本的降低在一定程度上被传统工作场所可能不需要的计算机和其他技能的需求所抵消。

4. 减少在制品库存和交货期

与作业车间环境相比，机电一体化系统中在制品的减少是相当显著的。据报道，有些设备的在制品减少了 80%，这可能是各种因素减少了零件等待金属切削操作时间的原因。这些因素包括：将生产零件所需的所有设备集中在一个小区域内；减少了所需的夹具数量；因为工序是在工作单元中组合的，所以应减少零件必须经过的机器数量，计算机有效调度零件批量进入机电一体化系统或在系统内调整。

5. 对不断变化的生产需求的响应能力

机电一体化具有固有的灵活性，可以随着市场需求的变化或工程设计的变化制造不同的产品。此外，所需的备件生产可以混合到正常运行中，而不会严重干扰正常的机电一体化系统的生产活动。

6. 维持生产的能力

许多机电一体化系统被设计成在一台或多台机器出现故障时能够合理地降级。这是通过采用冗余加工能力和允许故障机器通过的物料处理系统来实现的。因此，吞吐量可以保持较低的速率。

7. 产品质量高

有时，机电一体化系统的一个优势被忽视了，特别是与没有联合到合作系统中的机器相比，即提高的产品质量。产品设计特点与生产能力的基本整合、高水平的自动化、夹具数量和机器访问数量的减少、设计更好的永久性夹具，更加关注零件与机器的调整，以及对零件/机器准线的更多关注都决定了良好的单个零件质量和从一个工件到另一个工件的卓越一致性，从而大幅降低了返工成本。

8. 操作的灵活性

操作的灵活性使生产效率明显提高。在一些使用机电一体化的工厂中，中班和晚班实际上是在无人看管的情况下运行的。这种几乎无人值守的操作模式目前还算是例外。然而，随着更好的传感器和计算机控制技术被开发出来，操作的灵活性应该会变得越来越普遍。其可以检测和处理意外问题，如工具损坏和局部流动堵塞。在这种操作模式下，检查、固定和维护可以在早班进行。

9. 能力的灵活性

通过对可用面积的正确规划，机电一体化系统在设计之初就可以满足小批量生产的需要；随着需求的增加，可以很容易地增加新机器以提供所需的额外产能。

Evaluate

任务名称		机电一体化概论	姓名	组别	班级	学号	日期	
		考核内容及评分标准		分值	自评	组评	师评	平均分
三维目标	知识	机电一体化的概念、在工业应用中的优缺点及其相关技术的相关知识		25分				
	技能	能阅读专业英语文章、翻译专用英语词汇		40分				
	素养	在学习过程中秉承科学精神、合作精神		35分				
加分项	收获（10分）	收获（借鉴、教训、改进等）：	你进步了吗？				加分	
			你帮助他人进步了吗？					
	问题（10分）	发现问题、分析问题、解决方法、创新之处等：					加分	
总结与反思							总分	

2.2 Single Chip Microcomputer

Text

With the tremendous advances made in automatic control and automated equipment, the knowledge of single chip microcomputer is very important. We'll introduce some key hardware as follows.

Central Processing Unit

The central processing unit (CPU) is the brain of the microcontrollers reading user's programs and executing the expected task as per instructions stored there. Its primary elements are an 8 bit arithmetic logic unit (ALU), accumulator (ACC), few more 8 bit registers, B-register, stack pointer (SP), program status word (PSW) and 16 bit registers, program counter (PC) and data pointer register (DPTR).

The ALU (ACC) performs arithmetic and logic functions on 8 bit input variables. Arithmetic operations include basic addition, subtraction, multiplication and division. Logical operations are and, or, exclusive or as well as rotate, clear, complement and etc. Apart from all the above, ALU is responsible in conditional, branching decisions, and provides a temporary place in data transfer operations within the device. B-register is mainly used in multiply and divide operations. During execution, B-register either keeps one of the two inputs and then retains a portion of the result. For other instructions, it can be used as another general purpose register. Program status word keeps the current status of the ALU in different bits.

Stack pointer is an 8 bit register. This pointer keeps track of memory space where the important register information is stored when the program flow gets into executing a subroutine. Normally SP is initialized to 07H after a device reset and grows up from the location 08H. The stack pointer is automatically incremented or decremented for all PUSH or POP instructions and for all subroutine calls and returns. Program counter is the 16 bit register giving address of next instruction to be executed during program execution and it always points to the program memory space. Data pointer register (DPTR) is another 16 bit addressing register that can be used to fetch any 8 bit data from the data memory space. When it is not being used for this purpose, it can be used as two 8 bit registers.

Fig. 2-23 shows the diagram of the 8051 core.

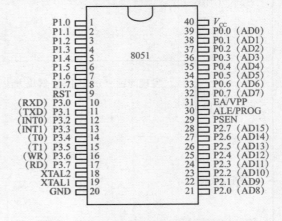

Fig. 2-23 The diagram of the 8051 core

Input/Output Ports

The 8051's in put/output (I/O) port structure is extremely versatile and flexible. The device has 32 I/O pins configured as four 8 bit parallel ports (P0, P1, P2 and P3). Each pin can be used as an input or as an output under the software control. These I/O pins can be accessed directly by memory instructions during program execution to get required flexibility. These port lines can be operated in different modes and all the pins can be made to do many different tasks apart from their regular I/O function executions. Instructions, which access external memory, use port P0 as a multiplexed address/data bus.

At the beginning of an external memory cycle, low order 8 bits of the address bus are output on P0. The same pins transfer data byte at the later stage of the instruction execution. Also, any instruction that accesses external program memory will output the higher order byte on P2 during the read cycle. Remaining ports, P1 and P3 are available for standard I/O functions. But all the 8 lines of P3 support special functions: two external interrupt lines, two read/write control lines, serial port's two data lines and two timer inputs are designed to use P3 port lines. When you don't use these special functions, you can use corresponding port lines as a standard I/O. Even within a single port, I/O operations may be combined in many ways. Different pins can be configured as inputs or outputs independent of each other or the same pin can be used as an input or output at different time.

【Words and phrases】

single chip microcomputer	单片机
microcontroller	n. 中央处理器；中央处理单元，[自]微控制器
execute	v. 执行；实施；处决；完成；表演（高难度动作）；制作（艺术作品）
instruction	n. 指令；命令；指示；教导；用法说明
logic	adj. 逻辑的
program status word (PSW)	程序状态字
program counter (PC)	程序计数器
stack pointer (SP)	堆栈指针
data pointer register (DPTR)	数据指针寄存器
program memory	程序寄存器

Course practice

1. Complex sentence analysis.

(1) This pointer keeps track of memory space where the important register

information is stored when the program flow gets into executing a subroutine.

① where the important register information is stored：由 where 引导的定语从句，where 在定语从句中作地点状语。

② when the program flow gets into executing a subroutine：是一个时间状语从句，when...，意思为当……时候。

【译文】当程序执行子程序时，指针记录重要寄存器信息的存储位置。

（2）The stack pointer is automatically incremented or decremented for all PUSH or POP instructions and for all subroutine calls and returns.

increment：n.＜正式＞（尤指连续、定量的）增长；v.（计算机）使（数字值）各自增加。

【译文】依照所有的入栈或出栈指令，堆栈指针自动增量或减量，依照子程序调用和返回指令，堆栈指针自动地调用和返回。

2．Translate the following paragraphs.

（1）The CPU is the brain of the microcontrollers reading user's programs and executing the expected task as per instructions stored there.

（2）Its primary elements are an 8 bit arithmetic logic unit（ALU），accumulator（ACC），few more 8 bit registers，B-register，stack pointer（SP），program status word（PSW）and 16 bit registers，program counter（PC）and data pointer register（DPTR）.

（3）The ALU performs arithmetic and logic functions on 8 bit input variables. Arithmetic operations include basic addition，subtraction，multiplication and division.

3．Answer to the questions according to the text.

（1）Which component reads user's programs and executes the expected task as per instructions stored there?

（2）What are composed of the basic elements of CPU?

（3）What is the function of ALU?

（4）What is the main task of B-register?

（5）Which ports are available for standard I/O functions generally?

Translation of the text

单片机

随着自动控制、自动化设备的快速发展，单片机知识显得非常重要。下面将介绍单片机的主要硬件。

中央处理器

中央处理器（CPU）是微处理器的大脑，它读取用户程序并根据存储在程序中的每条指令执行预期的任务。它的主要元件是一个 8 位运算逻辑单元（ALU）、累加器（ACC）、一些 8 位寄存器、变址寄存器、堆栈指针（SP）、程序状态字（PSW）和 16 位寄存器、程序计数器（PC）和数据指针寄存器（DPTR）。

ALU（ACC）对 8 位输入变量执行算术和逻辑功能。算术运算包括基本的加法、减法、乘法和除法。逻辑操作包括与、或、异或、循环、清零、补等。除此之外，ALU 还负责条件、分支选择，并在其内部提供一个临时存储空间以备数据传输操作。变址寄存器主要用于乘除法操作。乘除法操作中，变址寄存器先取两个输入数据中的一个运算，然后分别存放结果部分。对于其他指令，它可以作为另一个通用寄存器使用。程序状态字在寄存器的不同位保留 ALU 的当前状态。

堆栈指针是一个 8 位寄存器。当程序执行子程序时，指针记录重要寄存器信息的存储位置。一般而言，系统复位后，SP 通常初始化为 07H 并从 08H 单元开始增长。依照所有的入栈或出栈指令，堆栈指针自动增量或减量，依照子程序调用和返回指令，堆栈指针自动地调用和返回。程序计数器（PC）是一个 16 位寄存器，在执行程序时提供下一条指令的地址，保证执行下段指令，它总是指向程序存储空间。数据指针寄存器（DPTR）也是一个 16 位寄存器，可以从数据存储空间中读取任何一个 8 位数。当它不用于此目的时，可以用作两个 8 位寄存器。

8051 主要部件组成如图 2-23 所示。

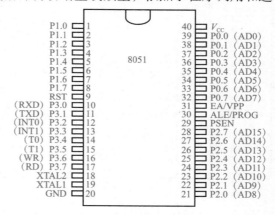

图 2-23　8051 主要部件组成示意

输入 / 输出接口

8051 的输入 / 输出接口结构是非常多且灵活的。它有 32 个输入 / 输出端子，配置为四个八位并行接口（P0、P1、P2 和 P3）。在软件控制下，每个端子线都可以用作输入或输出端口。这些输入 / 输出端口可以直接在程序执行中由存储指令灵活地直接访问。这些接口线能够在不同模式下运行，除常规的 I/O 指令外，它们可以分别完成很多不同的任务。访问外存储器的指令用 P0 接口作为多元化的地址 / 数据总线。

在一个外部存储周期开始时，地址总线的低 8 位在 P0 输出。同样，端口在指令执行的后期传递数据字节。并且在读取数据过程中，访问外存储器的指令在 P2 端口输出高 8 位。剩余的接口，P1 和 P3 作为标准输入 / 输出端口使用。但所有 P3 的 8 条线支持特殊功能：两条外部中断线、两条读写控制线、两个串行数据口和两个时间输入，被设置使用 P3 接口。当不使用这些特殊功能时，能使用相应的端口作为标准输入 / 输出端口。即使在一个端口内，输入 / 输出操作也可以多种方式组合。不同的端子可以设定成为彼此独立的输入或输出端口，或同一个端子在不同时段分别用作输入或输出端口。

Evaluate

任务名称		单片机		姓名		组别	班级	学号	日期
考核内容及评分标准						分值	自评	组评	师评	平均分
三维目标	知识	单片机、中央处理器、堆栈指针等相关知识				25 分				
	技能	能阅读专业英语文章、翻译专用英语词汇				40 分				
	素养	在学习过程中秉承科学精神、合作精神				35 分				
加分项	收获（10 分）	收获（借鉴、教训、改进等）：				你进步了吗？			加分	
						你帮助他人进步了吗？				
	问题（10 分）	发现问题、分析问题、解决方法、创新之处等：							加分	
总结与反思									总分	

2.3 Programmable Logic Controller

Text

A programmable logic controller, or programmable controller is a small computer used for automation of real-world processes, such as control of machinery on factory assembly lines. The PLC usually uses a microprocessor. The program can often control complex sequencing and is often written by engineers. The program is stored in battery backed memory and/or EEPROMs.

A PLC has the basic structure, as shown in Fig. 2-24.

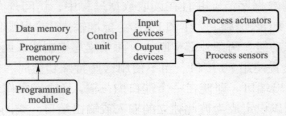

Fig. 2-24　Basic structure of a PLC

From Fig. 2-24, the PLC has four main units: the programme memory, the data memory, the output devices and the input devices. The programme memory is used for storing the instructions for the logical control sequence. The status of switches, interlocks, past values of data and other working data is stored in the data memory.

The main differences from other computers are the special input/output arrangements. These connect the PLC to sensors and actuators. PLCs read limit switches, temperature indicators and the positions of complex positioning systems. Some even use machine vision. On the process actuator side, PLCs drive any kind of electric motor, pneumatic or hydraulic cylinders or diaphragms, magnetic relays or solenoids. The I/O arrangements may be built into a simple PLC, or the PLC may have external I/O modules attached to a proprietary computer network that plugs into the PLC.

PLCs were invented as less expensive replacements for older automated systems that would use hundreds or thousands of relays and cam-timers. Often, a single PLC can be programmed to replace thousands of relays. Programmable controllers were initially adopted by the automotive manufacturing industry, where software revision replaced the rewiring of hard wired control panels.

The functionality of the PLC has evolved over the years to include typical relay control, sophisticated motion control, process control, distributed control systems and complex networking.

The earliest PLCs expressed all decision making logic in simple ladder logic inspired from the electrical connection diagrams. The electricians were quite able to trace out circuit problems with schematic diagrams using ladder logic. This was chosen mainly to reduce the apprehension of the existing technicians. Today, the PLC has been proven very reliable, but the programmable computer still has a way to go. According to the standard *Programming Industrial Automation Systems*: *Concepts and Programming Languages*, *Requirements for Programming Systems*, *Decision-Making Aids* (IEC 61131-3) issued by International Electro Technical Commission, it is now possible to program these devices using structured programming languages, and logic elementary operations.

A graphical programming notation called sequential function charts is available on certain programmable controllers.

However, it should be noted that PLCs no longer have a very high cost, typical of a "generic" solution. There are other ways for automating machines, such as a custom microcontroller based design, but there are differences among both. PLCs contain everything needed to handle high power loads right out of the box, while a microcontroller would needs power supplies, power modules, etc; also a microcontroller based design would not have the flexibility of in-field programmability of a PLC. That is why PLCs are used in production lines, for example. These typically are highly customized systems so the cost of a PLC is low compared to the cost of contacting a designer for a specific, one time only design. On the other hand, in the case of mass produced goods, customized control systems quickly pay for themselves due to the lower cost of the components.

【 Words and phrases 】

battery backed memory and/or EEPROM	随机存储器和（或）电可擦写可编程只读存储器
pneumatic	adj. 气动的；气力的
cylinder	n. 圆柱体；圆筒；汽缸
	n. 气胎
relay	n. [电工]继电器
solenoid	n. [电]螺线管
automate	adj. 自动化的
	v. 使自动化
cam	n. 凸轮
initially	adv. 最初；开头
hard wired	adj. 硬线连接的
ladder	n. 梯子；梯形
inspire	vt. 作为……的原因或根源；引起；激励；鼓励；激发行动
trace out	促动；描绘出
circuit	n. 电路
apprehension	n. 担心；忧虑
IEC	国际电工委员会
sequential function chart	顺序功能图
load	n. 负载；重荷
	v. 装上；装进；承载；装载
power supply	电源
module	n. 模数；模块
flexibility	n. 弹性；适应性；柔韧性

Course practice

1. Complex sentence analysis.

（1）The program is stored in battery backed memory and/or EEPROMs.

battery backed memory and/or EEPROMs：随机存储器和（或）电可擦写可编程只读存储器。

【译文】程序存储在随机存储器和（或）电可擦写可编程只读存储器中。

（2）PLCs were invented as less expensive replacements for older automated systems that would use hundreds or thousands of relays and cam-timers.

that would use hundreds or thousands of relays and cam-timers：这是一个定语从句，修饰 older automated systems；that 在定语从句中作主语，指 older automated systems；

定语从句可以翻译成 older automated systems 的定语。

【译文】PLC 替代使用成百上千个继电器和凸轮计时器的自动控制系统,更加实惠。

(3) However, it should be noted that PLCs no longer have a very high cost, typical of a generic solution.

① it should be noted that...: 值得注意的是……

② 这是一个主语从句,it 是形式主语,that PLCs no longer have a very high cost, typical of a generic solution 才是真正的主语,that 只是引导词,在主语从句中不充当任何成分。

【译文】然而,值得注意的是,PLC 不再是价格非常高的产品,而是一种通用的解决方案。

(4) Modem PLCs with full capabilities are available for a few hundred USD.

① available: adj. 可用的,可获得的;有空的,有闲暇的;未婚的。

② ...be available for...: ……可供……使用。

③ USD: United States dollar, 美元。

【译文】现在,功能完善的 PLC 也仅有几百美元。

2. Translate the following paragraphs.

(1) An example of a relay in a simple control application is shown in the figure. In this system the first relay on the left is used as normally closed, and will allow current to flow until a voltage is applied to the input A.

(2) The second relay is normally open and will not allow current to flow until a voltage is applied to the input B.

3. Fill in the blanks and translate the sentences.

(1) Fill in the blanks according to the words given in Chinese.

1) The development of low cost (   )(计算机) has brought the most recent resolution, the (   )(可编程逻辑控制器).

2) The advent of the PLCs began in the 1970s, and (   )(梯形图) is the main (   )(编程) method used for PLCs and has been developed to mimic (   )(继电器) logic.

3) The relays allow (   )(电源) to be switched on and off without a mechanical switch.

(2) Place a "T" after the sentence which is true and an "F" after the sentence which is false.

1) The instructions for the logical control sequence are stored in the data memory. (   )

2) The I/O modules can only be built into PLCs. (   )

3) Now, we can use structured programming languages and logic elementary operations to program. (   )

4) The simple ladder logic inspired from the electrical connection diagrams.（   ）

5) The programmability of a PLC is flexible.（   ）

Translation of the text

可编程逻辑控制器

可编程逻辑控制器，或者称可编程控制器是一种小型计算机，用于现实世界的自动化生产，如控制工厂的装配机械等。PLC通常使用微处理器，其程序可控制复杂的加工工序，通常由工程师编写。程序存储在随机存储器和（或）电可擦写可编程只读存储器中。

PLC的基本结构如图2-24所示。

图2-24 PLC的基本结构

由图2-24可见，PLC有四个主要的部件：程序存储器、数据存储器、输出装置和输入装置。程序存储器用来存储逻辑控制序列指令。开关状态、联锁状态、数据初始值和其他工作数据存储在数据存储器中。

与其他计算机比较，最大的区别是PLC有专用的输入/输出装置。这些装置将PLC与传感器和执行机构相连。PLC能读取限位开关、温度指示器和复杂定位系统的位置信息，有些甚至使用了机器视觉。过程执行机构方面，PLC能驱动任何一种电动机、气压或液压装置、控光装置、电磁继电器或螺线管。输入/输出装置安装在简单的PLC内部，或者PLC通过外部I/O模块连接到插入PLC的专用计算机网络上。

PLC替代使用成百上千个继电器和凸轮计时器的自动控制系统，更加实惠。通常一台PLC可以通过编程来替代数千个继电器。可编程控制器最初在自动化的制造工业中使用，通过修改软件可以取代硬线连接控制面板的重新连线。

PLC经过数年的发展，其功能包括典型的继电器控制、复杂的运动控制、过程控制、分布控制和复杂的网络技术。

最早的PLC采用简单的梯形图来描述其逻辑程序，梯形图源自电气连线图。电气技师使用逻辑图很容易找出原理图中的电路问题，这样在很大程度上减少了他们的担心。如今，PLC的功能已经非常可靠，但是可编程计算机仍然有其发展空间。按照国际电工委员会标准*Programming Industrial Automation Systems:Concepts and Programming Languages*，*Requirements for Programming Systems*，*Decision-Making Aids*（IEC 61131-3），现在可以使用结构化的编程语言和逻辑初等变换来进行编程。

某些可编程控制器可使用一种图表式的编程符号，这种符号称为顺序功能图表。

然而，值得注意的是，PLC 不再是价格非常高的产品，而是一种通用的解决方案。使机床实现自动化还有其他方法，如传统的基于微控制器的设计，但是这两者之间有一些差别。PLC 包含所有能直接操作大功率负载的输出功能，而微控制器还需要专用的电源和电源模块等；同时微控制器设计的方式不具备 PLC 现场可编程的灵活性。这也正是 PLC 在生产线上得到运用的原因之一。这些都是典型的高度用户化的系统，比起一次请一个设计人员做一个专门设计的成本，一台 PLC 的成本要低一些。另外，在大批量生产中，用户化系统能很快收回成本，因为它的构件成本低。

Evaluate

任务名称		可编程逻辑控制器	姓名	组别	班级	学号	日期	
		考核内容及评分标准		分值	自评	组评	师评	平均分
三维目标	知识	PLC 的概念及其功能		25 分				
	技能	能阅读专业英语文章、翻译专用英语词汇		40 分				
	素养	在学习过程中秉承科学精神、合作精神		35 分				
加分项	收获（10 分）	收获（借鉴、教训、改进等）：	你进步了吗？		加分			
			你帮助他人进步了吗？					
	问题（10 分）	发现问题、分析问题、解决方法、创新之处等：			加分			
总结与反思					总分			

2.4 Sensor Technology

Text

A sensor is a device, which responds to an input quantity by generating a functionally related output usually in the form of an electrical or optical signal. Sensors and sensor systems perform a diversity of sensing functions allowing the acquisition, capture, communication, processing, and distribution of information about the states of physical systems. This may be chemical composition, texture and morphology, large-scale structure, position, and so on. Few products and services of the modem society would be possible without sensors.

Sensor Value Chain

Sensor technology is distinctly interdisciplinary. Few organizations have all the competencies necessary for the realization of a sensor solution in-house.

The realization of a sensor product requires tasks to be completed, ranging all the way from product definition to final product and subsequent marketing and service.

Wireless Sensor Technology

For applications within health care, industrial automation, consumer products, and security there is a strong and growing need for wireless, self-powered sensors. Radio frequency identification technology (RFID) is an example of an emerging application with great potential. Sensors with wireless connections and no internal power supply are anticipated to become of great importance in areas like health care, consumer products, structural health monitoring, and energy tapping, etc.

A sensor field where a large number of connected sensor nodes are embedded. Each node will consist of a wireless sensor often without any internal power supply. The sensor interacts with a transceiver which is again connected to an infrastructure, possibly to a so-called sink. The collection of data may be controlled by a managing device.

Biometric Sensors

This is another area where some markets are expected to exhibit strong growth over the next years. Fingerprint identification equipment and iris scanners are examples of such markets that are spurred by the increasing demand within security.

Non-invasive & Non-contact Sensors

An increasing number of applications call for non-contact sensing. Light and sound play important roles both independently and combined. Here, combinations of ultrasound and light are expected to become important in order to overcome the limitations inherent when using either light or sound independently.

Miniaturization and Integration

Sensors are often used in large production plants. However, online sensing often has to be done in areas with limited space. Hostile environments require robust sensors and robustness may be obtained by miniaturization and integration. In optics, this may imply new system and fibre optics. In acoustics, new non-contact methods for excitation are being devised.

Novel Materials

New materials are needed if a number of very different functional requirements are to be fulfilled at sufficiently low cost. It may be systems combining microfluidics with light generation and detection. Polymers are anticipated to play an increasingly larger role, both due to potential low cost and due to great flexibility in functional properties. Recent development of advanced microscopes has made it possible to see and even move single atoms and molecules. This opens opportunities for creating entirely new materials and processes. The technology has become known as nanotechnology and currently receives a lot of attention.

Not many companies have thus far reached the stage of commercialization, but intense research is in progress and expectations are very high.

Sensor Fusion and Sensor Networks

Complex systems will often be monitored by a number of very different types of sensors. X-ray can reveal properties of weldings online, optics can detect chemical composition and macroscopic dynamics, whereas ultrasound may provide information about the inner structures of systems.

Sensor fusion is combining such multisensory information in order to obtain new functionality. Moreover, systems with a large number of low-cost sensors coupled in networks are becoming increasingly important.

【Words and phrases】

sensor technology	传感技术
quantity	n. 数目；数量；参量
diversity	n. 多样性；不同；差异
capture	n. 捕获；占领；存储
	v. 俘获；记录；占领；体现
processing	n. 过程语言（一种开源编程语言）
	v. 处理；加工；审核；计算
chemical composition	化学成分
texture	n. 质地；纹理；口感
	v. 使……具有浮凸特征
morphology	n. 形态学；词法；结构
the sensor value chain	传感器价值链
distinctly	adv. 清楚地；明显地；非常
interdisciplinary	adj. 跨学科的
competency	n. 能力；本领；权限；感受态；反应能力
realization	n. 认识；领悟；实现；变卖；体现；发声
range from... to...	范围从……到……
wireless sensor technology	无线传感技术
security	n. 保护措施；安全工作；保安部门；保障；安全感
radio frequency identification technology（RFID）	无线射频识别技术
potential	adj. 潜在的；可能的
	n. 潜能；电压；电位
transceiver	n. [通信] 接收器
infrastructure	n. 基础设施
so-called sink	所谓的水槽

biometric sensor	生物识别传感器
non-invasive & non-contact sensor	非插入式和非接触式传感器
ultrasound	n. 超声；超声波
miniaturization	n. 小型化；微型化
integration	n. 整合；一体化；融合
fibre optic	光纤
microfluidic	n. 微流体
polymer	n. 聚合物
nanotechnology	n. 纳米技术
commercialization	n. 商业化
fusion	n. 融合；结合；核聚变；熔化；融合物
detect	v. 查明；察觉；测出；检测；识别
macroscopic dynamics	宏观动力学

 Course practice

1. Complex sentence analysis.

（1）Sensor technology is distinctly interdisciplinary.
　interdisciplinary：*adj.* 跨学科的。
【译文】传感器技术具有明显的跨学科性。

（2）Polymers are anticipated to play an increasingly larger role, both due to potential low cost and due to great flexibility in functional properties.
　be anticipated to do：被期望/预计用于做某事。
【译文】聚合物由于其潜在的低成本和在功能特性上的巨大灵活性，预计将发挥越来越重要的作用。

（3）Sensor fusion is combining such multisensory information in order to obtain new functionality.
　multisensory：*adj.* 多种感觉的；使用多种感觉器官的。
【译文】传感器融合将这些多感官信息结合起来，以获得新的功能。

2. Match the items listed in the following two columns.

1.	指纹识别	a.	fingerprint identification
2.	卫生保健	b.	macroscopic
3.	消费品	c.	ultrasonic
4.	能源开发	d.	health care
5.	纳米技术	e.	energy exploitation
6.	超声波	f.	nanotechnology technology
7.	宏观的	g.	microscopic
8.	微观的	h.	consumer product

3. Translate the following paragraphs.

(1) A biosensor is a device for the detection of an analyte that combines a biological component with a physicochemical detector component. A canary in a cage, as used by miners to warn of gas could be considered a biosensor.

(2) Many of today, biosensor applications are similar, in that they use organisms which respond to toxic substances at a much lower level than us to warn us of their presence. Such devices can be used both in environmental monitoring and in water treatment facilities.

4. Choose the proper answer to fill in the blank and translate the sentences.

(of, to, at, in, on)

(1) Each node will consist (   ) a wireless sensor often without any internal power supply.

(2) Combinations of ultrasound and light are expected (   ) become important in order to overcome the limitations inherent.

Translation of the text

传感器技术

传感器是一种能够响应输入量，将其转换成以电或光信号输出的相关功能的装置。传感器和传感器系统能执行各种各样的检测功能，并能够采集、获取、传递、处理和分配物理系统的状态信息。这些信息可能是化学成分、晶体结构和构造、大型结构、位置等。现代社会几乎没有一种产品和设备不用传感器。

传感器的价值链

传感器技术具有明显的跨学科性。很少有机构拥有能实现传感器全部解决方案所需的能力。

一种传感器产品的推出，需要完成从产品定义、产品最终成型到后续的市场营销和服务等一系列的工作。

无线传感技术

卫生保健、工业自动化、消费品和安全等领域，对无线、自供电的传感器的需求不断增长。射频识别技术（RFID）便是一个实例，并显示了其巨大的应用潜力。无线连接和没有内部电源的传感器将在卫生保健、消费品、结构健康监测和能源发掘等领域发挥重要的作用。

传感器中嵌入了大量连接的传感器节点。每一个节点都由无线传感器构成，通常不含内部电源。这些传感器与无线收发器相连，而收发器与其下属元器件相连，有可能与一个所谓的接收器相连。数据的收集由管理系统控制。

生物传感器

生物传感器领域的市场有望在随后的几年中呈现很强的增长势头。例如，指纹识别装置和虹膜扫描设备的市场便是受不断提高的安全需求的刺激而发展起来的。

非插入式和非接触式传感器

越来越多的科学应用要求使用非接触式传感器。光和声音正在这个领域独立或共同发挥着很重要的作用。在此,为了克服光或声音单独使用时固有的局限性,超声波和光的组合运用正变得越来越重要。

小型化和集成化

在大型的生产车间经常要用到传感器。然而,在线检测只能在有限的空间进行。恶劣的环境需要优质的传感器,而其优良的性能可以通过小型化和集成化来获得。在光学方面,这意味着新的系统和光纤的使用。在声学方面,正在研发新的非接触式声激励装置。

新型材料

如果要以极低的成本实现多种完全不同的功能,就要用到新型材料。它可能是一些结合使用了光生成和探测技术的微流控技术的系统。聚合物由于其潜在的低成本和在功能特性上的巨大灵活性,预计将发挥越来越重要的作用。新近开发的先进显微镜已经能够看到甚至移动单个原子和分子,这为人们创造全新的材料和加工工艺提供了机遇。通常所说的纳米技术,目前已受到了广泛的关注。

能够将这一领域的发展达到工业化程度的公司目前还不多,很多都只是在进行积极的研究工作,并寄予很高的期望。

传感器整合和传感器网络

复杂的系统经常是用大量不同类型的传感器进行监控的。X 光可以在线显示焊接的质量,光学可检测化学合成和宏观动力学特性,而超声波能提供反映系统内部构造的信息。

传感器融合将这多感官信息结合起来,以获得新的功能。此外,使用大量低成本的传感器组成的网络系统也变得越来越重要。

Evaluate

任务名称		传感器技术		姓名	组别	班级	学号	日期
考核内容及评分标准				分值	自评	组评	师评	平均分
三维目标	知识	传感器、无线传感技术、生物传感器等基本概念		25 分				
	技能	能阅读专业英语文章、翻译专用英语词汇		40 分				
	素养	在学习过程中秉承科学精神、合作精神		35 分				
加分项	收获(10分)	收获(借鉴、教训、改进等):		你进步了吗?			加分	
				你帮助他人进步了吗?				
	问题(10分)	发现问题、分析问题、解决方法、创新之处等:					加分	
总结与反思							总分	

2.5 CAD/CAM

 Text

If you are involved with CNC machines, you are probably involved with CNC programming and CAD/CAM systems. CAD (computer aided design) is widely used to describe any software capable of defining a mechanical component with geometry, surfaces, or solid models. CAM (computer aided manufacturing) is software used to develop CNC program.

Engineering design and manufacturing uses CAD/CAM software for three distinctly different purposes.

Design Modeling

A mechanical design engineer uses CAD software to create a part. This definition of the part can be called its model. This model can be represented as a drawing or a CAD data file.

A manufacturing engineer or CNC programmer, uses CAD software to:

(1) Develop a computer model of a part defined by a drawing;

(2) Evaluate and repair the design CAD data to manufacturing tolerances. This is a surprisingly common task.

Create new part models from the original design to allow for manufacturability. This would include adding draft angles or developing models of the part for different steps in multi-process manufacturing.

Design models of fixtures, mold cavities, mold cores, mold bases, and other tooling. An example of manufacturing modeling needs to be done before a part can be machined.

CNC Program

A manufacturing engineer or CNC programmer uses CAM software to select tools, methods, and procedures to machine the models defined in the manufacturing modeling section described above.

In the beginning, there were only CAD systems. Engineers only used CAD systems to draw pictures of parts. The first CAM systems helped a CNC programmer/machinist/manufacturing engineer program from these drawings. This making of drawings, and programming parts from drawings, was (and still is) time consuming and subject a lot of human errors. Someone got the bright idea to eliminate this to-and-from drawing step, and integrated CAD/CAM was born.

Machining Methods

You can cut imported CAD models in two ways. The first one is to generate the tool path from the surfaces and tool geometry, applying edge protection to the surfaces, or generate the tool path directly from the solid, using Boolean operators and accounting for the sweep volume of the tool.

The second requires lots of calculations based on the intersection of the solids, because it has to be continuously calculated as the tool moves along. Because the math is complex, risk of errors is high and the probability of gouging is much increased. It is common to generate 10,000 to 100,000 individual tool path moves, and an error in a single move stops machining altogether. Any compromise gives a bad finish.

No software or hardware wizardry can speed up CNC programming if the CAM system uses an inefficient technique to generate tool paths. The most robust method tessellates the part into polyhedrons consisting of numerous triangles (facets or planes), where user-defined accuracy determines the number of facets. Finished part accuracy is typically between 0.000 1–0.000 5 in (0.002 54–0.012 7 mm), so the math solution to the intersection of tool shape and facets is quicker and more reliable than the formulas to generate tool paths directly from solid.

It's easier to plan, organize and reorder cutting routines if your CAM system includes a module that lets you graphically cut and paste into an efficient operations manager. This sort of feature lets you view and record CNC parameters associated with tool paths, and provides a history of machining strategies for family of parts programming. Tool, material and step-over settings used for one CNC operation can be applied to different geometry for new tool paths generation.

Good math is only half of what a CAM system is about. The other half is good machining practice. Following are some features and methods for machining.

Gouge-Free Tool Support

Today's CAM systems should offer support for square-nosed, bull-nosed and ball-nosed end mills, with full gouge avoidance on each. Because each tool advantages in machining various shapes, the better systems do not limit the operator's choice of tools. Rough cutting with bull-nosed tools is faster and, depending on the workpiece, can also be effective for finish cutting.

Surface Edge Protection

Individual shapes that make up a free-form surface or solid model may have gaps, and they're handled differently among CAM systems. Some will simply add a straight line between the surfaces on either end of the gap, but this risks gouging. The more reliable method is to "edge protect" each of the individual surfaces. With this method the tool path is guaranteed not to gouge the interior or edges of the surface.

Tool Path Bounding

The extent of the tool path should be limited to cut only a portion of the model. For quick tool path generation, the bounding needs to be done at tool path creation time rather than waiting until after the full tool path is generated. This is especially significant if a portion of the model is all you want to cut anyway.

CAM software should allow the user to specify a fixed distance between each cut, and specify the maximum scallop height and maximum surface tolerance. Then the software will calculate the proper step-over to develop the shortest possible CNC code and fastest machining cycles.

Feed Between Moves

Tool path generation usually follows rows as it cuts the part. When the tool comes to the end of a row, the CAM system decides how to position the tool at the start of the next row. Some use a simple "up and over" algorithm, but, watch out, this can gouge if the software isn't check first. To avoid gouging, better system will offer the option "following the part in the transition to the start of the next row".

Inspect Gouge in Engage and Leave

The engage move positions the machine tool to the start of the first cutting row. The machinist needs a variety of options to engage and to leave the part gently, such as spiral, arc, and ramp motions, to guarantee they won't gouge.

Cutting Methods

Cutting techniques for both solid and surface models are Z-level roughing (contour line), Z-level (contour line) finishing, planar cutting, single surface flow, multi-surface flow cutting (includes radial cutting), and project cutting. The machinist should have the option to specify various cutting methods for any portion of the model.

Z-level roughing is the most efficient way to remove the majority of excess material which slices a model with constant Z-level planes to generate spiral or zigzag paths. If there are overhanging areas on the part, the CAM system can avoid gouging with the shank of the tool as well as the tip.

Z-level finishing is the best way to finish areas of a part that are almost vertical. It is important that the operator be able to apply a boundary curve to limit the area of path generation.

Planar cutting is the oldest and most well known method of cutting complex models. It allows the operator to generate parallel planar tool paths over any number of underlying surfaces. Current systems vary widely in the speed and quality of tool paths generated. Even systems that claim to generate gouge-free tool paths may gouge on complex shapes. If you are choosing a new CAM system, be sure to test it on the most complex part you're likely to cut.

Single-surface flow cutting leaves the best surface when a part's geometry is defined by only a few surfaces. It cuts only a single surface at a time, but the CAM system must be capable of gouge checking that surface against any number of surrounding surfaces.

Multi-surface flow cutting is a powerful feature. For machining a boomerang-shaped part, or any part with a bend or a lot of bends, it is important that the CAM system can generate tool paths that flow along or across a specified flow surface while remaining essentially perpendicular to it. On a round, oval or odd-shaped part, mold or die, this feature lets you cut perpendicular to shape around the part's periphery, regardless of its shape. The machinist can control tool-path direction to produce the finest detail.

Projected cutting is useful when machining an irregularly shaped part with a particular 2D tool path, as for engraving. CAM system should be able to create this tool path, project it onto the complex model and cut-all with full gouge avoidance.

【Words and phrases】

Boolean operator	布尔运算
wizardry	n. 巫术；魔法
tessellate	adj. 镶嵌细工的；嵌成花纹的；
	vt. 镶嵌
module	n. (机器、建筑物等的)组件；模件； (课程的)单元；模块
graphic	n. (商业设计或插图中的)图形；图样；图案制图学；制图法
	v. 用图案装饰；给……加上花样；仿照；模仿
	vt. 使……呈球形；包围
gouge	vt. 用半圆凿子挖；欺骗
	n. 沟；圆凿；以圆凿刨
square-nosed end mill (cylindrical end mill)	方头立铣刀(圆柱立铣刀)
bull-nosed end mill	牛头立铣刀
scallop height	残留高度
slice	n. (切下的食物)薄片；片
	v. 把……切成薄片；切；割；划

Course practice

1. Complex sentence analysis.

(1) A manufacturing engineer or CNC programmer uses CAM software to select tools, methods, and procedures to machine the models defined in the manufacturing modeling section described above.

【译文】制造工程师或数控编程员使用 CAM 软件选择刀具、加工方法和加工工序加工上述定义的制造模型。上述为直译，改为意译："为了加工上述定义的制造模型，制造工程师或数控编程员使用 CAM 软件选择刀具、加工方法和加工工序"，这符合汉语的表达习惯。

(2) It is important that the operator be able to apply a boundary curve to limit the

area of path generation.

该句为虚拟语气，故 that 后从句谓语系动词为原形 be，也可为 should be。

【译文】操作员能够运用边界曲线限制刀具轨迹的生成区域是很重要的。

2. Translate the following paragraphs.

（1）Someone got the bright idea to eliminate this to-and-from drawing step, and integrated CAD/CAM was born.

（2）A manufacturing engineer or CNC programmer uses CAD software to evaluate and repair the design CAD data to manufacturing tolerances.

（3）Creating the model is only the first step in developing a customer's new part. In order to bring the part to production, a versatile CAM system is needed to import the model and convert the surfaces for machining.

（4）It's common to generate 10,000 to 100,000 individual tool path moves, and an error in a single move stops machining altogether. Any compromise gives a bad finish.

（5）CAD/CAM is a unified software system, in which the CAD portion is interfaced inside the computer with the CAM system.

（6）However, in the near future some companies will not use drawings at all, but will be passing part information directly from design to manufacturing via a database.

3. Answer the following questions according to the text.

（1）How do you usually save the drawing?

（2）For what purposes do engineering design and manufacturing use CAD/CAM software?

（3）What cutting methods do today's CAM systems offer?

（4）Which CAM software do you use?

 Translation of the text

CAD/CAM

如果你从事与 CNC 机床有关的工作，很可能会与 CNC 和 CAD/CAM 系统打交道。CAD（computer aided design，计算机辅助设计）是指各种能够用几何模型、曲面模型或实体模型定义机械零件的软件。CAM（computer aided manufacturing，计算机辅助制造）是用于开发 CNC 程序的软件。

工程设计和制造使用 CAD/CAM 软件有三个截然不同的目的。

设计建模

机械设计工程师使用 CAD 软件创建零件。这里的零件可以定义为模型，模型可用图纸或 CAD 数据文件表示。

制造工程师或 CNC 编程员使用 CAD 软件进行以下工作。

（1）创建由图纸定义的零件的计算机模型。

（2）评估和修正 CAD 设计数据，以符合制造公差要求。这是最为常见的工作。

根据原始设计建立适合制造的新零件模型，这些工作包括加入拔模角度或者开发多工序制造中不同工序的零件模具。

设计夹具、模腔、模芯、模架和其他加工所需的模具。零件在加工前需制造模样。

数控编程

为了加工上述定义的制造模型，制造工程师或数控编程员使用 CAM 软件选择刀具、加工方法和加工工序。

起初只有 CAD 系统，工程师只使用 CAD 系统进行零件绘图。早期的 CAM 系统协助 CNC 编程员 / 机械师 / 制造工程师根据这些 CAD 图纸进行编程。这种制图方法以及根据零件图纸进行编程的方法仍然是十分费时、费力的，并且容易受人为错误因素的影响。有人想出了一个好主意，消除了这个来回绘图的步骤，于是集成CAD/CAM 就诞生了。

加工方法

可用两种方法加工输入的 CAD 模型：第一种方法运用曲面边界保护，由曲面和刀具几何形状生成刀具轨迹；运用布尔运算，计算刀具的扫掠体，直接由实体模型生成刀具轨迹。

第二种方法需要进行大量的计算，因为刀具的移动需要连续的基于实体相交点的计算值。由于所用的数学算法非常复杂，所以产生错误的可能性很大，并且加工时产生过切的可能性也大幅增加。通常生成 10 000 ~ 100 000 次单独的刀具轨迹移动，只要有一次出错，就中断了整个实体加工过程，这都会影响产品表面光洁度。

如果 CAM 系统使用效率很低的方法生成刀具轨迹，任何软件或硬件向导都不能提高 CNC 编程的速度。最好的方法是将零件分成由无数三角形（小平面或平面）组成的多面体，用户指定的精度决定了小平面的数目。成品件的精度通常为 0.000 1 ~ 0.000 5 in（0.002 54 ~ 0.012 7 mm），因此用数学求解刀尖形状和小平面的交点比用公式直接由实体生成刀具轨迹更加快速、可靠。

如果 CAM 系统有能够进行图形化剪切和粘贴操作的管理器模块，则规划、组织以及重新安排切削路线会容易得多。这种功能可以查看、记录与刀具轨迹有关的 CNC 参数，并且能提供同族零件编程的加工路线的历史记录。而且一次 CNC 操作中的刀具、材料和步距设置可应用于生成不同的新刀具轨迹。

正确的数学计算只是 CAM 系统的一半，另一半是合理的加工实践。以下是一些加工特点和方法。

无过切刀具支持

现在的 CAM 系统应该支持平头立铣刀、牛头铣刀和球头铣刀，而且完全能避免过切。因为每种刀具在加工不同形状时各有优点，所以较好的 CAM 系统不限制操作者对刀具的选择。用牛头铣刀进行粗加工更快，对于有的工件，用牛头铣刀进行精加工也很有效。

曲面边界保护

构成自由体曲面或实体模型的多个单独形体之间会有间隙，不同的 CAM 系统对间隙会有不同的处理方法。有的系统只是在曲面间隙两端加一条直线，但这很可能产生过切。较可靠的方法是对各单独曲面进行"边界保护"。采用这种方法，保证了刀具轨迹不会对曲面内部或边沿造成过切。

刀具轨迹边界

为了只切削模型的一小部分，要限制刀具轨迹的范围。为了快速生成刀具轨迹，必须在刀具轨迹生成时划分区域，而不是等到生成完整的刀具轨迹后才划分区域。如果只是切除模型的一小部分，这种方法尤其重要。

CAM 软件应该允许用户指定每个切削步距，并指定最大残留高度和最大表面粗糙度公差，然后软件将计算合理的步距，生成最短的 CNC 代码和最快的加工循环。

两个切削路径之间的转移

刀具切削零件时，刀具轨迹的生成通常按行进行。刀具移动到行的末端时，CAM 系统确定如何将刀具定位到下一行的起点。一些系统使用简单的"抬刀横越"算法，但是必须注意，如果不首先进行软件检查，这很可能产生过切。为了避免过切，良好的系统能提供"跟踪零件形状，转移到下一行起点"的选择功能。

切入和退刀时的过切检查

切入动作将机床定位到第一切削行的起点。机械师需要通过各种选择功能温和地切入和退出工件，如螺旋式进退刀、圆弧式进退刀以及斜坡式进退刀，以保证不产生过切。

切削方法

实体模型和曲面模型的加工方法包括 Z 水平（等高线）粗加工、Z 水平（等高线）精加工、平面切削、单曲面加工、多曲面加工（包括径向加工）和投影加工。机械师应该能根据模型的形状选择不同的切削方法。

Z 水平粗加工是最有效的切除零件大部分多余材料的切削方法。它通过恒定的 Z 水平平面将模型分层，产生跟随轮廓或往复式的刀具轨迹。如果零件上有凸出区域，CAM 系统能避免由于刀杆和刀尖产生的过切。

Z 水平精加工是完成陡坡零件精加工的最好方法。操作员能够运用边界曲线限制刀具轨迹的生成区域是很重要的。

平面切削是应用最早、最有名的复杂模型切削方法。采用这种方法，操作员可以在下底面上生成多层的平行平面刀具轨迹。目前很多系统在刀具轨迹生成的速度和质量上差别很大。声称能生成无过切刀具轨迹的系统在加工复杂形状时也可能会产生过切。如果要选择新的 CAM 系统，那么一定要在尽可能复杂的零件上进行试切削。

当零件的几何形状由为数不多的几个曲面定义时，采用单曲面加工得到的零件表面质量最好。虽然一次只切削一个面，但是 CAM 系统必须能够检查该曲面是否会与任意周围曲面产生过切。

多曲面加工具有强大的功能。要加工飞镖形状的零件或有一处或多处弯曲的零件，CAM 系统必须能够生成沿着或穿过指定表面的刀具轨迹，并且刀具与曲面保持基本垂直，这个功能非常重要。在圆形、椭圆形或不规则形状零件和模具的加工上，不管零件的形状如何，该功能都可以围绕零件外侧，与零件外形垂直进行加工。机械师能够控制刀具轨迹的方向，从而制造出最精密的产品。

当用 2D 刀具轨迹加工不规则形状的零件时，投影加工最有用，如雕刻。CAM 系统应能生成该刀具轨迹，并将其投射到复杂的模型上，然后进行切削，所有这一切都不产生过切。

Evaluate

任务名称		CAD/CAM		姓名	组别	班级	学号	日期
		考核内容及评分标准			分值	自评	组评	师评	平均分
三维目标	知识	CAD/CAM 相关的设计建模、数控编程、加工方法等相关知识		25 分				
	技能	能阅读专业英语文章、翻译专用英语词汇		40 分				
	素养	在学习过程中秉承科学精神、合作精神		35 分				
加分项	收获（10 分）	收获（借鉴、教训、改进等）：	你进步了吗？		加分	
			你帮助他人进步了吗？			
	问题（10 分）	发现问题、分析问题、解决方法、创新之处等：		加分		
总结与反思				总分		

2.6 CAE/CIM

 Text

I. CAE

Computer aided engineering (CAE) made it possible to achieve a decisive improvement in mold design, to optimize the process, and above all, to improve and speed up the design of injection molds and progressive die design.

Simulation Software Simplifies Injection Mold Design

C-MOLD, as one of simulation software, is a set of integrated CAE simulations for plastics molding processes, including injection molding filling, post-filling, and cooling; part shrinkage and warpage; co-injection molding; gas-assisted injection molding; reactive molding; blow molding; thermoforming; injection/compression molding and microchip encapsulation. C-MOLD provides solutions in all stages of design and manufacturing, to improve productivity and enhance part quality. Key features include a motif graphical user interface; a sophisticated geometry modeler and finite-element mesher tailored to plastics applications; seamless integration with major CAD and structural CAE applications; a polymer database containing over 11,000 materials; CAE analyses that provide critical information for optimizing part, mold, and process design in plastics molding operations; and an easy-to-use data visualizer for viewing analysis results.

Simulation Software Simplifies Progressive Die Design

In this era of global warp speed and virtual reality, calculating the deep draws of progressive dies or the spring-back of metal is performed by simulation software instead of the earlier trial-and-error method. These software programs essentially replace the artistic methods of die-making with scientific formulas. For more than a decade tool engineers have been designing parts, progressive die sequences, and dies and detailing cutter paths as a continuous process using computer aided design.

Manufacturing Process

Progressive die design used to be considered as just a road map for the die-maker during the build process. First the form stations were cut and given to the assembly department to be mounted on the die shoes. Later the die was taken to the press, where tryout began. This process was labor-intensive and required many hours of press time.

Today, die-makers begin to make molds by using part simulations instead of using network maps. In the process of mold forming, there are deep drawing process, tipping process and shaping process. This process is evaluated and determined to produce no break and no wrinkle parts by using two tools in the software, the forming limit diagram and the material thickness change map.

The FLD shows points plotted on two curves that are created from the material properties. All points should fall into the area below the first curve. Points above the first curve and below the second are in the marginal zone, areas of the part that are prone to fractures. For parts in this zone, any variation in the process—such as material property changes, thickness changes, or even a change in lubrication—could cause part failure. Points above the second curve are fractures in the part.

Designing Carrier Webs

Today, a web is taken out of the CAD design and set up in the simulation with locator pins in the pilot holes. The locator pins lock that area of the web so that it will not move during the simulated press stroke.

The remaining webs react by stretching as they will be in the real press, which shows where clearances are needed in steels for the movement of the stretch webs. They must have adequate stretch to allow the part to form, but enough strength to advance the part through the tool.

With the development of today's forming simulations, the blanking and trimming stations are developed virtually. The finished-part geometry is projected onto the simulation result, and the software reverses the forming process, producing a part that becomes what is cut for the final trim steels.

Virtual tryout allows the whole tool to go directly into machining and assembly. With this process, the die-maker knows that the die can produce the part when it is in the press. The die-maker then adjusts the part to bring it into tolerance.

Under Pressure

Simulation software can accurately calculate how much pressure will be required to hold the part for each of the forming stations, thus allowing an accurate estimate of the number of nitrogen cylinders that are required. This is important because if the pressure is too low, the binder may gap, which might cause uncontrolled material flow, thinning, or wrinkles in the part. On flange stations, if the pressure is incorrect, the center of the part may bow upward, which causes spring-back in the flange walls.

Reducing Tryout Time

While simulation software will not tell the die-maker how to build the tooling, it will accurately predict the result of the chosen process and help the designer develop a robust process to produce the part.

When coupled with a solid modeling CAD program, the software allows a virtual construction and tryout of the die in the computer before any steel is sent to the CNC machining center. This allows for the complete assembly of 95% of all tools before they make the first hit, greatly reducing the amount of tryout time.

II. CIM

Computer integrated manufacturing (CIM) is the term used to describe the most modern approach to manufacturing. Although CIM encompasses many of the other advanced manufacturing technologies such as CNC, CAD/CAM, robotics and just-in-time (JIT) delivery, it is more than a new technology or a new concept. Computer integrated manufacturing is actually an entirely new approach to manufacturing or a new way of doing business.

To understand CIM, it is necessary to begin with a comparison of modem and traditional manufacturing. Modern manufacturing encompasses all of the activities and processes necessary to convert raw materials into finished products, deliver them to the market and support them in the fields. These activities include the following:

(1) Identifying a need for a product;
(2) Designing a product to meet the need;
(3) Obtaining the raw materials needed to produce the product;
(4) Applying appropriate processes to transform the raw materials into finished products;
(5) Transporting product to the market;
(6) Maintaining the product to ensure a proper performance in the field.

This broad, modern view of manufacturing can be compared with the more limited traditional view that focuses almost entirely on the conversion processes. The old approach separates such critical pre-conversion elements as market analysis research, development and design for manufacturing, as well as such after-conversion elements as product delivery and product maintenance. In other word, in the old approach to manufacturing, only those processes that take place on the shop floor are considered manufacturing. This traditional approach of separating the overall concept into numerous specialized elements alone was not fundamentally changed with the advent of automation. While the separate elements themselves became automated (i.e., computer aided drafting and design and CNC in machining). They remained separate. Automation alone did not result in the integration of these islands of automation.

With CIM, not only are the various elements automated, but also the islands of automation are all linked together or integrated. Integration means that a system can provide complete and instantaneous sharing of information. In modern manufacturing, integration is accomplished by background computers. With this CIM can now be defined as the total integration of all manufacturing elements through the use of computers.

Furthermore, such an illustration can show that research, development, design, marketing, sales, shipping, receiving, management and pre-execution personnel all have instant access to all the information generated in this system. This is exactly the inside meaning of a CIM system. Progress is being made toward the eventual realization of CIM in manufacturing. When this is accomplished, fully integrated manufacturing firms will realize a number of benefits from CIM:

(1) Product quality increases;
(2) Lead time is reduced;
(3) Direct labor cost is reduced;

(4) Product development time is reduced;
(5) Inventories are reduced;
(6) Overall productivity increases;
(7) Design quality increases.

【Words and phrases】

I

computer aided engineering (CAE)	计算机辅助工程
decisive	*adj.* 决定性的
optimize	*v.* 使最优化
above all	最重要；首先
speed up	加速
progressive die	级进模具：一种用于金属冲压加工的工具
injection molding filling	注塑成型填充
post-filling	保压
cooling	*n.* 冷却
part shrinkage and warpage	制品收缩与翘曲
co-injection molding	共注成型
gas-assisted injection molding	气辅注射成型
reactive molding	反应成型
thermoforming	*n.* 热成型
injection/compression molding	压注成型
microchip encapsulation	微芯片封装
key feature	主要特点
motif graphical	功能图形
user interface	用户界面
geometry modeler	几何图形建模器
finite-element mesher	有限元网格划分器（模块）；剖分（模块）
visualizer	*n.* 数据可视化（功能模块）
warp speed	高速飞驰
road map	路线图
forming limit diagram (FLD)	成型极限图
fall into	位于
be prone to	易于
locator pin	定位销
pilot hole	定位孔

be projected onto	投射在
be coupled with	加上

II

computer integrated manufacturing (CIM)	计算机集成制造
encompass	v. 包围；环绕
just-in-time (JIT)	即时
delivery	n. 递送；交付
more than… or…	与其说……还不如说……
convert	v. 转变；变换
raw material	原料
finished product	成品；产品
identify	v. 认出；识别
appropriate	adj. 适当的
maintain	v. 维持；保持
ensure	v. 保证
performance	n. 履行；执行
be compared with…	与……比较
product maintenance	产品维护
on the shop floor	车间一线；工作现场
advent	n. 出现；到来
integrated	adj. 整合的；完整的
instantaneous	adj. 瞬间的；即刻的
modern manufacturing	现代制造业
background	n. 背景；后台
accomplish	v. 实现；完成

Course practice

1. Complex sentence analysis.

(1) Computer aided engineering (CAE) made it possible to achieve a decisive improvement in mold design, to optimize the process, and above all, to improve and speed up the design of injection molds and progressive die design.

① make it possible to do…译为使做……成为可能。

② it：为形式宾语，真正的宾语是 to achieve a decisive improvement in mold design, to optimize the process, and above all, to improve and speed up the design of injection molds。

【译文】计算机辅助工程（CAE）使在模具设计方面取得决定性的进步成为可能，使工艺过程最优化，尤其是改进和加速了注塑模具和级进模具设计。

(2) Simulation software can accurately calculate how much pressure will be required

to hold the part for each of the forming stations, thus allowing an accurate estimate of the number of nitrogen cylinders that are required.

本句为宾语从句，句中 calculate 后接从句 how much pressure will be required to hold the part for each of the forming stations，而由 thus 引导的结构为伴随状语，说明前面的部分。

【译文】仿真软件能够准确地计算出每一个成型工位上支撑零件所需的压力大小，从而可以准确地估计出所需氮气缸的数量。

（3）Although CIM encompasses many of the other advanced manufacturing technologies such as CNC, CAD/CAM, robotics and just-in-time (JIT) delivery, it is more than a new technology or a new concept.

more than... or...：大（多、甚）……或……，不止（只，仅）……或……，不只是……或……。

【译文】虽然 CIM 包括许多其他的先进制造技术，如计算机数控（CNC）、CAD/CAM、机器人技术和即时（JIT）交货等，但它不仅仅是一种新技术或新概念。

（4）Modern manufacturing encompasses all of the activities and processes necessary to convert raw materials into finished products, deliver them to the marke and support them in the fields.

本句属于 necessary to do... 句型，为形容词短语作后置定语修饰 activities and processes，意思是做……所必需的；convert... into... 译为把……转换成……。

【译文】现代制造包括把原材料转换成产品、把产品投向市场和进行技术支持所必需的一切活动和过程。

（5）This traditional approach of separating the overall concept into numerous specialized elements alone was not fundamentally changed with the advent of automation.

动名词短语 separating the overall...与介词一起构成后置定语修饰 approach。

【译文】这种将整体概念专门划分为许多专门元素的传统方法并没有随着自动化的出现而从根本上改变。

2. Match the following terms or phrases.

（1）computer aided engineering (CAE)　　　注射成型填充
（2）computer aided design (CAD)　　　　　吹塑成型
（3）computer aided manufacturing (CAM)　　压注成型
（4）geometry modeler　　　　　　　　　　功能图形
（5）injection molding filling　　　　　　　　计算机辅助制造
（6）co-injection molding　　　　　　　　　几何图形建模器
（7）blow molding　　　　　　　　　　　　材料（性能）数据库
（8）injection/compression molding　　　　　用户界面
（9）motif graphical　　　　　　　　　　　计算机辅助工程
（10）polymer database　　　　　　　　　　共注成型
（11）user interface　　　　　　　　　　　计算机辅助设计

3. Translate the following sentences.

(1) Polymer processing involves the transformation of a solid (sometimes liquid) polymeric resin, which is in a random form (e.g., powder, pellets, beads), to a solid plastics product of specified shape, dimensions, and properties.

(2) Injection molding, a discontinuous process, is extremely complex and is dependent on material properties, product and mold geometry as well as on process parameters.

(3) C-MOLD, as one of simulation software, is a set of integrated CAE simulations for plastics molding processes.

(4) It is not only the major manufacturers of raw materials who are offering simulation as a service to support their sales efforts; engineering consulting offices are also offering this service, as a consequence, simulation is now gaining a foothold in small and medium-sized businesses.

4. Answer the following questions.

(1) Why polymer processing simulation has been used to control the process and product quality?

(2) How many software suppliers are there? What polymer processing can be simulated by the software of C-MOLD?

(3) Calculation software for processes (e.g., C-PITA) have been tested and adapted to molding, haven't they?

(4) What is the CIM?

(5) What is the relationship between CIM and CNC, CAD/CAM, JIT?

(6) What is the difference between modern and traditional manufacturing?

(7) What does the word integration mean in this text?

(8) Can you tell us some benefits from CIM?

5. Mark the following statements with T(true) or F(false) according to tne text.

(1) CIM is a new type of enterprise. (　　)

(2) CIM includes CNC, CAD/CAM and JIT delivery. (　　)

(3) Modern manufacturing encompasses all of the activities to convert raw materials into finished products. (　　)

(4) The broad, modern view of manufacturing focuses almost entirely on the conversion processes. (　　)

(5) In the modern approach to manufacturing, only those processes that take place on the shop floor are considered manufacturing. (　　)

(6) CIM can now be defined as the total integration of all manufacturing elements through the use of computers. (　　)

Translation of the text

CAE/CIM

Ⅰ．CAE

计算机辅助工程（computer aided engineering，CAE）使在模具设计方面取得决定性的进步成为可能，使工艺过程最优化，尤其是改进和加速了注塑模具和级进模具设计。

仿真软件简化了注塑模具设计

作为一种模拟软件，C-MOLD 是一套用于塑料成型工艺的 CAE 模拟软件的集成，包括注塑成型填充、保压、冷却；制品收缩与翘曲；共注成型；气辅注射成型；反应成型；吹塑成型；热成型；压注成型和微芯片封装。C-MOLD 提供了从设计到制造各阶段的解决方案，以提高生产率并提升产品质量。其主要性能包括功能图形用户界面；为塑料应用而定制的完善的几何图形建模器和有限元网格划分器；具有主要 CAD 软件和结构化 CAE 软件应用的无缝集成；含有 11 000 多种材料的材料数据库；在塑料成型操作中为优化零件、模具和工艺设计提供重要信息的 CAE 分析；可看到分析结果的易于使用的数据可视化工具（功能模块）。

仿真软件简化了级进模具设计

在这个全球高速发展和虚拟现实的时代，用仿真软件计算级进模具的拉伸量和金属的可弹量，取代了早期的反复试验法。这些软件基本上用科学公式取代了制造模具的精湛方法。10 多年来，机械工程师一直在用计算机辅助设计将零件设计、级进模具工艺顺序、模具设计，甚至细化到刀具路径设计作为一个连续的流程来做。

制造过程

级进模具的设计过去常常被视为只是模具制造者在形成工艺过程中的一个路线图。首先模具在第一个工位上被加工并送到装配部门将其安装在模座上，接下来模具被送到压床上开始调试。这一过程是劳动密集型的，需要许多时间。

如今，模具制造者开始通过运用零件仿真模拟来制造模具而不再利用网络图。在模具成型过程中执行仿真的工序有拉深工序、翻边工序和整形工序。通过使用软件中的两个工具即成型极限图和料厚变化图来评估和决定此过程是否可以生产出无断裂和无起皱的零件。

FLD 表示出了由材料性能产生的两条曲线上的图形点。所有点都应位于第一条曲线以下的区域，第一条曲线以上和第二条曲线以下的点都处于边缘区，即零件易断裂的区域。对于此区域内零件，在加工过程中的任何变化，如材料属性的变化、厚度的变化，甚至是润滑的变化都会导致零件损坏。第二条曲线以上的点都处于零件的断裂区。

过程网络设计

如今，网络已经从 CAD 设计中取消了，建立在带定位孔和定位销的仿真模拟中。定位销锁住网络区，使其在仿真压机冲程过程中不会移动。

当网络位于真实压床上时，保留的网络会通过延伸做出反应，这一过程表明模具在该处需要为网络的延伸留有间隙。支撑板必须有足够的延伸量，才有助于零件加工成型，但是足够的强度又会使零件穿过模具。

随着当今成型仿真的发展，落料和修边工序得到了实质的发展。成品件的几何形状在仿真结果中被预测出来，此软件的仿真结果是成型过程的反过程，仿真生产了能够体现所用钢的最终修整的零件。

虚拟使用过程允许整个模具直接进入加工和装配过程。在这个过程中，模具制造者知道当模具位于压床上时，它可以生产零件。然后模具制造者调整零件在合适的公差范围内。

承受压力

仿真软件能够准确地计算出每一个成型工位上支撑零件所需的压力大小，从而可以准确地估计出所需氮气缸的数量。这一点很重要，因为如果压力过低，压料面可能产生间隙，这样在零件中可能产生不可控的材料流动、变薄或起皱。在翻边工序，如果压力不正确，零件的中心可能会向上弯曲，这样会使翻边回弹。

减少调试时间

虽然仿真软件不会告诉模具制造者如何制造模具，但是它会准确地预测所选工艺的结果，并帮助设计者制定一个较好的生产零件的工艺。

加上实体建模的 CAD 程序，该软件可以在把任何钢材送到 CNC 加工中心之前在计算机中虚拟构造并使用模具，这就使 95% 的模具在第一次冲压前完成了全部装配，这就大幅减少了调试时间。

Ⅱ．CIM

计算机集成制造（CIM）是用来描述最现代的制造方法的专业术语。虽然 CIM 包括许多其他先进制造技术，如 CNC、CAD/CAM、机器人技术和即时（JIT）交货等，但它不仅仅是一种新技术或新概念。计算机集成制造实际上是一种全新的制造方式，也可以说是一种全新的经营手段。

为了理解 CIM 的概念，有必要将现代制造和传统制造做一个比较。现代制造包括把原材料转换成产品、把产品投向市场和进行技术支持所必需的一切活动和过程。这些活动包括如下内容：

（1）确定产品需求；
（2）设计满足需求的产品；
（3）获取制造产品所需要的原材料；
（4）采用合适的方法把原材料转换成产品；
（5）把产品输向市场；
（6）维护产品以确保其具备符合规格的性能。

这种广泛的现代制造业观点可以与更有限的传统观点相比，传统观点几乎完全关注转换过程。旧的方法将市场分析研究等关键的转换前的要素和为生产而进行的开发和设计，以及产品交付和产品维护等转换后的要素分开。换句话说，在旧的制造方法中，只有那些在车间进行的过程才被认为是制造。这种将整体概念单独划分为许多专门元素的传统方法并没有随着自动化的出现而从根本上改变。虽然独立的元素本身变得自动化（即计算机辅助绘图和设计以及 CNC 加工）。但它们仍然是独立的。自动化本身并没有导致这些自动化孤岛的整合。

使用 CIM，不仅可以实现各种元素的自动化，而且可以将自动化孤岛连接在一起或集成在一起。集成意味着系统可以提供完整和即时的信息共享。在现代制造业中，集成是由后台计算机完成的。这种 CIM，现在可以定义为通过使用计算机对所有制造要素进行全面集成。

此外，这样的说明可以表明，研究、开发、设计、市场、销售、出货、收货、管理和采购人员都可以即时访问该系统中产生的所有信息，这正是 CIM 系统的内在含义。在制造业中最终实现 CIM 方面正在取得进展。当实现这一目标时，完全集成的制造公司将从 CIM 中获得许多好处：

（1）提高产品质量；
（2）缩短交货期；
（3）降低直接人工成本；
（4）减少产品开发时间；
（5）减少库存；
（6）提高整体生产效率；
（7）提高设计质量。

Evaluate

任务名称		CAE/CIM	姓名	组别	班级	学号	日期
		考核内容及评分标准	分值	自评	组评	师评	平均分
三维目标	知识	与 CAE/CIM 相关的 C-MOLD、级进模具、成型极限图等相关知识	25 分				
	技能	能阅读专业英语文章、翻译专用英语词汇	40 分				
	素养	在学习过程中秉承科学精神、合作精神	35 分				
加分项	收获（10 分）	收获（借鉴、教训、改进等）：	你进步了吗？			加分	
			你帮助他人进步了吗？				
	问题（10 分）	发现问题、分析问题、解决方法、创新之处等：				加分	
总结与反思						总分	

Module 3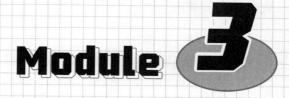

Typical Electromechanical Industry

Focus

- Unit 1　Refrigeration & Air-Conditioning
- Unit 2　Civil Aviation Maintenance
- Unit 3　Oil Industry

Unit 1　Refrigeration & Air-Conditioning

Unit objectives

【Knowledge goals】
1. 掌握显热储存、潜热储存等相关知识。
2. 掌握老虎钳、熔丝拔钳、螺栓、扳手等概念。
3. 掌握换热器结构、分类及其功能等相关知识。

【Skill goals】
1. 能够阅读专业英语文章。
2. 可以翻译专用英语词汇。

【Quality goals】
1. 在学习过程中发扬科学研究精神。
2. 增强团队合作意识。

1.1　Thermal Storage

制冷与空调

Text

Thermal storage is the temporary storage of high or low temperature energy for later use. Examples of thermal storage are the storage of solar energy for night heating and the storage of summer heat for winter use, the storage of winter ice for space cooling in the summer, and the storage of heat or coolness generated electrically during off-peak hours for use during subsequent peak hours. Most thermal storage applications involve a 24-hours storage cycle, although weekly and seasonal cycles are also used.

When the period of energy availability is longer than the period of energy use, thermal storage permits the installation of smaller heating or cooling equipment than would otherwise be required. In many cases, storage can provide benefits for both heating and cooling, either simultaneously or at different times of the year.

Conditions favoring thermal storage include high loads of relatively short duration; high electric power demand charges; low cost electrical energy during off-peak hours; need for cooling backup in case of power outage or a refrigeration plant failure (only chilled water pumps must be powered by the emergency generator); need to provide cooling for small after-hour loads such as cooling of restaurants; mid-height elevator equipment rooms, individual offices, and computer rooms; building expansion (installation of storage may eliminate the need for heating/cooling plant expansion); need to provide a fire fighting reservoir; need to supplement a limited capacity generation plant; and others.

In sensible heat storage, storage is accomplished by raising or lowering the temperature of the storage medium, i.e., water, rock beds, bricks, sand, or soil.

In latent heat storage, storage is accomplished by a change in the physical state of the storage medium, usually from liquid to solid (heat of fusion) or vice versa. Typical materials are water/ice, salt hydrates, and certain polymers. Energy densities for latent heat storage are greater than those for sensible heat storage, which results in smaller and lighter storage devices and lower storage losses.

Storage efficiency is the ratio of energy that can be withdrawn from storage, divided by the amount put into storage. Storage efficiencies up to 90% can be achieved in well-stratified water tanks that are fully charged and discharged on a daily cycle. All storage devices are subject to standby losses, which are generally proportional to the exposed surface area of the storage vessel.

For heating, ventilation and air conditioning (HVAC) and refrigeration purposes, water and phase change materials (PCMs) particularly ice-constitute the principal storage media. Soil, rock, and other solids are also used. Water has the advantage of universal availability, low cost, and transportability through other system components. Some phase change materials are viscous and corrosive and must be segregated within the container for the heat transfer medium. If heating and cooling storage is required, two phase change materials must be provided, unless heat pumping is used.

【 Words and phrases 】

thermal	*adj.* 热的；热量的
temporary	*adj.* 暂时的；临时的；临时性的
solar	*adj.* 太阳的；和太阳有关的
availability	*n.* 可用性；可得性；空闲
simultaneously	*adv.* 同时地
backup	*n. & vt.* 后援；支持；阻塞；做备份
emergency	*n.* 紧急情况；突发事件；非常时刻，紧急事件
reservoir	*n.* 水库；蓄水池
accomplish	*vt.* 完成；达到；实现
hydrate	*n. & v.* 水合物；与水化合
density	*n.* 稠密；浓厚
withdraw	*vt.* 收回；撤销
standby	*n.* 可以信任的人；待命的信息；备用品
vessel	*n.* 船；容器；器皿；脉管；导管
phase	*n.* 阶段；状态；相；相位
transportability	*n.* 可运输性；应处以流放
component	*n.* 元（部、组）件；成（部）分

viscous　　　　　　　　　　　　adj. 黏性的；黏滞的；黏稠的
corrosive　　　　　　　　　　　adj. 腐蚀的；蚀坏的；腐蚀性的
segregate　　　　　　　　　　　vt. 分开（离），对……实行种族隔离

Course practice

1. **Complex sentence analysis.**

(1) Thermal storage is the temporary storage of high or low temperature energy for later use.

thermal storage：热储存。

【译文】热储存暂时储存高温或低温能量，以备日后使用。

(2) Examples of thermal storage are the storage of solar energy for night heating and the storage of summer heat for winter use, the storage of winter ice for space cooling in the summer, and the storage of heat or coolness generated electrically during off-peak hours for use during subsequent peak hours.

off-peak hours：离峰时间（电力）。非交通繁忙时间；非峰荷时间。例如：

Collect this periodically during peak and off-peak hours. 在高峰和非高峰时段定期收集这些信息。

That meter will charge different prices for electricity depending on whether it's drawn during peak or off-peak hours. 该电表可以区别高峰时段和非高峰时段以提供不同的电价。

【译文】热储存的例子有储存太阳能用于夜间供热，储存夏季的热量供冬季使用，储存冬季的冰用于夏季的空间降温，以及在非用电高峰时段储存电能产生的热量或冷量，以供在随后的高峰时段使用。

(3) Conditions favoring thermal storage include high loads of relatively short duration; high electric power demand charges; low cost electrical energy during off-peak hours; need for cooling backup in case of power outage or a refrigeration plant failure (only chilled water pumps must be powered by the emergency generator); need to provide cooling for small after-hour loads such as cooling of restaurants; mid-height elevator equipment rooms, individual offices, and computer rooms; building expansion (installation of storage may eliminate the need for heating/cooling plant expansion); need to provide a fire fighting; need to supplement a limited capacity generation plant; and others.

①这是一个长句。在科技英语中经常会遇到很长的句子，要认真分析其成分。本句是一个简单句，主语是 conditions，favoring thermal storage 是现在分词作定语修饰 conditions；谓语是 include，后面的是宾语部分，由若干并列成分构成。

② power outage：停电。

③ after-hour：下班以后。

【译文】有利于热储存应用的条件包括：相对短的时间内有较高的载荷；高电力的加载需求；在非高峰时段的低成本电能；需蓄冷以防停电或制冷设备出现故障（仅仅是冷水泵需要以应急发电机驱动）；需要为下班后的小负荷提供制冷，如餐馆制冷；中等的电梯设备房间、个人办公室，以及计算机房间；建筑物的扩建（储存设备的安装可以不需要扩建加热/冷却机器）；提供消防水池的需求；需要补充一个小容量的发电机；等等。

（4）Storage efficiency is the ratio of energy that can be withdrawn from storage, divided by the amount put into storage.

① the ratio of：比值。例如：

And in practical terms, we can define the efficiency as the ratio of the heat into the workout. 在实际中，我们定义效率为热与功的比值。

② withdrawn：收回，撤销。例如：

The summons was withdrawn. 传票被撤回。

【译文】储存效率是指可以从储存中提取的能量与储存的能量的比值。

（5）All storage devices are subject to standby losses, which are generally proportional to the exposed surface area of the storage vessel.

句中的 which 作关系代词，引导非限制性定语从句，指代 losses。

【译文】所有的储存装置都会受到待机损耗的影响，而待机损耗通常与储存容器的暴露表面积成正比。

2. Translate the following sentences.

（1）When the period of energy availability is longer than the period of energy use, thermal storage permits the installation of smaller heating or cooling equipment than would otherwise be required.

（2）In many cases, storage can provide benefits for both heating and cooling, either simultaneously or at different times of the year.

 Translation of the text

热储存

热储存暂时储存高温或低温能量，以供日后使用。热储存的例子有储存太阳能用于夜间供热，储存夏季的热量供冬季使用，储存冬季的冰用于夏季的空间降温，以及在非用电高峰时段储存电能产生的热量或冷量，以供在随后的高峰时段使用。大多数热储存以24小时作为储存周期，但也有以一周和一个季度作为储存周期的。

当能量获取的周期比能量使用的周期长时，热储存允许使用比实际需要更小的加热或制冷设备。在许多情况下，储存能同时或在一年内不同时期为加热和制冷提供能量。

有利于热储存应用的条件包括：相对短的时间内有较高的载荷；高电力的加载需求；在非高峰时段的低成本电能；需蓄冷以防停电或制冷设备出现故障（仅仅是冷水

泵需要以应急发电机驱动）；需要为下班后的小负荷提供制冷，如餐馆制冷；中等的电梯设备房间、个人办公室，以及计算机房间；建筑物的扩建（储存设备的安装可以不需要扩建加热/冷却机器）；提供消防水池的需求；需要补充一个小容量的发电机；等等。

在显热储存中，储存是通过提高或降低储存介质，即水、岩床、砖、沙或土壤的温度来完成的。

在潜热储存中，储存是通过改变储存介质的物理状态来完成的，通常从液体变为固体（聚变热）或相反。典型的物质有水/冰、含盐的水合物和某些聚合物。潜热储存的能量密度比显热储存大，从而导致更小、更轻的储存装置和更低的储存损失。

储存效率是指可以从储存中提取的能量与储存的能量的比值。以一日为周期，在完全分层的水箱中充满水再放掉，可达到 90% 的储存效率。所有的储存装置都会受到待机损耗的影响，而待机损耗通常与储存容器的暴露表面积成正比。

对于供热通风与空气调节（HVAC）及制冷，水和相变材料（PCMs），特别是冰，组成主要的储存介质，另外，土壤、岩石及其他固体也被使用。水的优点是可以较容易获取，费用低，通过其他系统部件就能输送。一些相变材料是具有黏滞性和腐蚀性的，并且必须被隔离在用于存放传热介质的容器中。如果需要储热与储冷，除非使用热泵，否则必须提供两种相变材料。

Evaluate

任务名称		热储存	姓名	组别	班级	学号	日期
考核内容及评分标准			分值	自评	组评	师评	平均分
三维目标	知识	显热储存、潜热储存等相关知识	25 分				
	技能	能阅读专业英语文章、翻译专用英语词汇	40 分				
	素养	在学习过程中秉承科学精神、合作精神	35 分				
加分项	收获（10 分）	收获（借鉴、教训、改进等）：	你进步了吗？			加分	
			你帮助他人进步了吗？				
	问题（10 分）	发现问题、分析问题、解决方法、创新之处等：				加分	
总结与反思						总分	

1.2 Tools and Equipment

Text

The air-conditioning technician must work with electricity. It is necessary to use various tools and equipment safely. Special tools are needed to install and maintain electrical service to air-conditioning units. Wires and wiring should be installed according to *National Technical Specification for the Safety of Electric Equipments* (GB 19517—2023). Following is a brief discussion of the more important tools used by the electrician in the installation of air conditioning and refrigeration equipment.

Pliers

Pliers come in a number of sizes and shapes designed for special applications, see Fig. 3-1 (a). Pliers are available with either insulated or un insulated handles. Although pliers with insulated handles are always used when working on "hot" wires, they must not be considered sufficient protection alone. Other precautions must be taken. Long-nose pliers are used for close work in panels or boxes.

Fuse Pullers

Fuse pullers are designed to eliminate the danger of pulling and replacing cartridge fuses by hand in Fig. 3-1 (b). It is also used for bending fuse clips, adjusting loose cutout clips, and handling live electrical parts. It is made of a phenolic material, which is an insulator. Both ends of the puller are used. Keep in mind that one end is for large-diameter fuses; the other is for small-diameter fuses.

Screwdrivers

Screwdrivers come in many sizes and tip shapes. Those used by electricians and refrigeration technicians should have insulated handles. For safe and efficient use, screwdriver tips should be kept square and sharp. They should be selected to match the screw slot. See Fig. 3-1 (c).

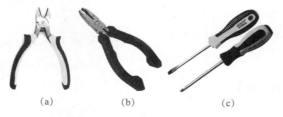

(a)　　　　　(b)　　　　　(c)

Fig. 3-1 Tools 1

(a) Plier; (b) Fuse puller; (c) Screwdriver

The Phillips-head screwdriver has a tip liking a star and is used with a Phillips screw. These screws are commonly found in production equipment. The presence of four slots assures that the screwdriver will not slip in the head of the screw.

Wrenches

Three types of wrenches used by the air-conditioning and refrigeration trade are shown in Fig. 3-2.

The adjustable open-end wrenches are commonly called crescent wrenches.

Monkey wrenches are used on hexagonal and square fittings such as machine bolts, hexagonal nuts, or conduit unions.

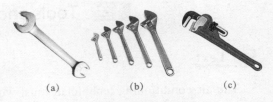

Fig. 3-2 Tools 2

(a) Fixed wrench; (b) Crescent wrenches; (c) Pipe wrench

Pipe wrenches are used for pipe and conduit work. They should not be used where crescent or monkey wrenches can be used. Their construction will not permit the application of heavy pressure on square or hexagonal material. Continued misuse of the tool in this manner will deform the teeth on the jaw face and mar the surfaces of the material being worked.

Soldering Equipment

The standard soldering equipment used by electricians consists of the same equipment that the refrigeration mechanics use, as shown in Fig. 3-3 (a). It consists of a non-electric soldering device in the form of a torch with propane fuel cylinder or an electric soldering iron, or both.

Tool Kits

Some tool manufacturers make up tool kits for the refrigeration. See Fig. 3-3 (b) for a good example. In the snap-on tool kit, the leak detector is part of the kit. The gages are also included. An adjustable wrench, tubing cutter, hacksaw, flaring tool, and ball-peen hammer can be hung on the wall and replaced when not in use. One of the problems for any repair person is keeping track of tools.

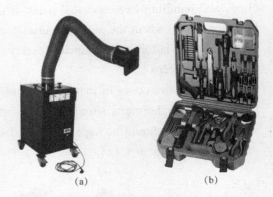

Fig. 3-3 Soldering equipment and tool kit

(a) A soldering equipment; (b) A tool kit

【Words and phrases】

plier	n. 钳子；老虎钳
fuse puller	熔丝拔钳
cartridge fuse	熔断丝管；熔丝管；筒式熔断器
phenolic	n. & adj. [胶粘剂] 酚醛树脂；[有化] 酚的
screwdriver	n. 螺丝刀；螺钉旋具

crescent	*n.* 新月；弦月；新月形的东西
hexagonal	*adj.* 六边的；六角形的
propane	*n.* ［有化］丙烷
hacksaw	*n.* 钢锯；可锯金属的弓形锯
	vt. 用钢锯锯断
flaring tool	［机］扩口工具；胀管器

Course practice

1. Complex sentence analysis.

（1）Following is a brief discussion of the more important tools used by the electrician in the installation of air conditioning and refrigeration equipment.

① 句中 following 是主语，brief discussion 是补语；介词短语 of the more important tools 修饰 discussion；used by 引导的过去分词短语修饰 tools。

② brief：简明扼要的。

例如：Now please be brief — my time is valuable. 现在请长话短说——我的时间很宝贵。

【译文】下面简单介绍一下电工在安装空调和制冷设备时使用的比较重要的工具。

（2）Pliers come in a number of sizes and shapes designed for special applications...

① 句中 pliers 是主语，come in 是谓语，a number of sizes and shapes 是宾语。过去分词短语 designed for special applications 修饰 sizes and shapes。

② come in：开始使用；起作用；出现。例如：

Nylon stockings came in soon after the end of the war. 战后不久尼龙袜子就时兴起来。

Here is the plan of attack, and this is where you come in. 这是进攻示意图，你负责这个地区。

That's where the trouble often comes in. 那就是经常出现纠纷之处。

【译文】老虎钳有多种尺寸和形状，专为特殊应用而设计……

（3）Their construction will not permit the application of heavy pressure on square or hexagonal material.

① 句中 construction 是主语，will not permit 是谓语，the application 是宾语。借此短语 of heavy pressure 和 on square or hexagonal material 用来修饰宾语。

② permit：许可，允许。例如：

Try to go out for a walk at lunchtime, if the weather permits. 如果天气条件允许，午饭时尽量出去散散步。

Smoking is not permitted. 不许吸烟。

【译文】它们的结构不允许在方形或六边形材料上施加重压。

（4）Continued misuse of the tool in this manner will deform the teeth on the jaw face and mar the surfaces of the material being worked.

①句中 continued misuse of the tool in this manner 做主语，后面是并列成分。will deform 和 mar 是谓语，the teeth 和 the surfaces 是宾语。

② misuse：滥用；误用。例如：

I hate to see her misusing her time like that. 我不愿看到她把时间用在不正当的事情上。

The misuse of power and privilege. 权力和特权的滥用。

【译文】继续以这种方式滥用工具，会使钳口牙齿变形，并且在材料的工作面上留下瑕疵。

2. Translate the following paragraph.

The Phillips-head screwdriver has a tip liking a star and is used with a Phillips screw. These screws are commonly found in production equipment. The presence of four slots assures that the screwdriver will not slip in the head of the screw.

 Translation of the text

工具与设备

空调技术员工作时必须用电，安全地使用各种工具和设备是必要的。在安装和维护空调装置的电气设备时需要使用专用工具，电线和线路应按《国家电气设备安全技术规范》（GB 19517—2023）安装。下面简单介绍一下电工在安装空调和制冷设备时使用的比较重要的工具。

老虎钳

老虎钳有多种尺寸和形状，专为特殊应用而设计，如图 3-1（a）所示。老虎钳可以使用绝缘或不绝缘的手柄，虽然老虎钳用于"火线"工作时带有绝缘手柄，但依然不能认为已经具备足够的防护，还必须有其他防护措施。在面板或箱内近距离作业时采用长头钳。

熔丝拔钳

熔丝拔钳用来排除手工拔下和更换熔丝筒的危险，如图 3-1（b）所示。它也可以用来弯曲熔丝、调整松动的保险盒和接触带电元件。它由绝缘的酚醛树脂材料制造而成。熔丝拔钳的两端都有用。注意，一端用于大号熔丝，另一端用于小号熔丝。

螺钉旋具

螺钉旋具有许多尺寸及刀头形状。电工和空调技术员用的螺钉旋具必须带有绝缘手柄。为了能安全有效地使用，螺钉旋具头必须保持方形且锋利。要选择不同的螺钉旋具以适用于不同螺钉的槽口，如图 3-1（c）所示。

十字头螺钉旋具的尖端呈星形，与十字头螺钉配合使用。十字头螺钉在生产设备上很常见，其带有四个插槽可确保螺钉旋具头不会滑出。

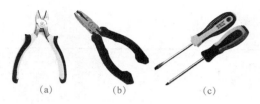

图 3-1 工具 1
(a) 老虎钳；(b) 熔丝拔钳；(c) 螺钉旋具

扳手

如图 3-2 所示为三种用于空调与制冷行业的扳手。

图 3-2 工具 2
(a) 固定扳手；(b) 可调扳手；(c) 管钳

可调节开口扳手通常称为钩扳手。

活动扳手用于六角形或方形配件，如机械螺栓、六角螺母或管子。

管钳用于管子和导管作业。在能用钩扳手和可调扳手的地方不必使用管钳。它们的结构不允许在方形或六边形材料上施加高压。继续以这种方式滥用管钳，会使钳口牙齿变形，并且在材料的工作面上留下瑕疵。

焊接装置

电工使用的标准焊接装置与制冷技工使用的设备相同，如图 3-3（a）所示。该装置包括不带电的丙烷喷枪或电烙铁，或者两者兼有。

工具箱

一些工具制造商为制冷设备生产工具箱。如图 3-3（b）所示为一款很好的工具箱。在这款按扣式的工具箱里，有检漏仪，还有量规。当不用可调节扳手、管子切割机、钢锯、扩口工具、圆头锤时，可挂在墙上，不使用时可进行调换。对于任何维修工来说，其中一个问题就是跟踪工具的情况。

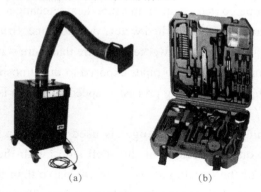

图 3-3 焊接装置与工具箱
(a) 焊接装置；(b) 工具箱

Evaluate

任务名称		工具与设备	姓名	组别	班级	学号	日期	
		考核内容及评分标准		分值	自评	组评	师评	平均分
三维目标	知识	老虎钳、熔丝拔钳、螺钉旋具、扳手等相关知识		25 分				
	技能	能阅读专业英语文章、翻译专用英语词汇		40 分				
	素养	在学习过程中秉承科学精神、合作精神		35 分				
加分项	收获（10 分）	收获（借鉴、教训、改进等）：	你进步了吗？			加分		
			你帮助他人进步了吗？					
	问题（10 分）	发现问题、分析问题、解决方法、创新之处等：				加分		
总结与反思						总分		

1.3 Heat Exchanger

Text

If a material is to be cooled, a shell-and-tube heat exchanger can also be used. The cooling medium is normally water, chilled water, or cold brine. Brine (a solution of salt in water) is used because it freezes at a lower temperature than pure water. Sometimes liquids are cooled by circulating them through pipes exposed to the atmosphere, this is called an air-cooled heat exchanger. The cooling process is speeded if a fan is used to blow cool air rapidly over the tubes (Fig. 3-4).

When a shell-and-tube heat exchanger is used for cooling, the material to be cooled may be either on the tubeside or the shell side, with the cooling medium on the other side. If one of the two liquids is more corrosive than the others (brine, for example, can be corrosive), it is usually on the tube side. The tubes can then be made of a corrosion-resistant material, with a cheaper material for the shell. If a corrosive material is used on the shell side—in contact with both the inner surface of the shell and the outer surface of the tubes—both the tubes and shell must be able to withstand corrosion(Fig. 3-5).

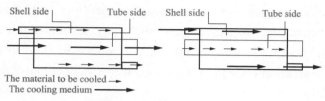

Fig. 3-4　Two forms of cooling

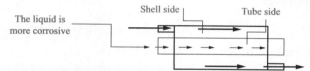

Fig. 3-5　Cooling form of corrosive liquids

It is possible to operate shell-and-tube heat exchangers in either the horizontal or vertical position, but horizontal exchangers are most common. When steam is used for heating, it is always introduced through the upper opening, or nozzle, in the shell, so that the lower opening can be used for draining condensate. If a pool of condensate builds up in such an exchanger(usually due to malfunctioning or wrongly sized steam trap) and submerges some of the tubes, the heat transfer suffers because condensing steam transfers heat more rapidly than does hot water. When a liquid is used on the shell side, it is always introduced through the lower nozzle and withdrawn through the upper nozzle. In this way the engineer can be assured that the exchanger remains full, with liquid surrounding all the tubes(Fig. 3-6, Fig. 3-7).

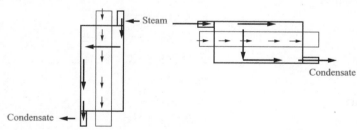

Fig. 3-6　Two types of shell-and-tube heat exchangers 1

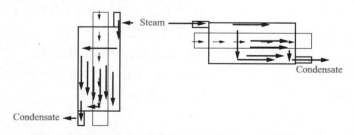

Fig. 3-7　Two types of shell-and-tube heat exchangers 2

Shell-and-tube heat exchangers are so widely used that they are available as stock items from many manufacturers. There are a number of standard designs used throughout the industry. However, no group of standard exchangers could fulfill all needs, so many exchangers are custom designed and manufactured. The size of the shell-and-tube heat exchanger needed for any particular job depends mainly on the amount of heat to be transferred and the volume of fluid to be handled. The calculations are learned in college by most engineers, but computer programs now do them automatically.

【Words and phrases】

shell	*n.* 贝壳；壳；外形；炮弹
shell-and-tube heat exchanger	管壳式换热器
medium	*n. & adj.* 媒体；方法；媒介；中间的；中等的
chill	*n. & adj. & v.* 寒冷的；使冷；变冷；冷藏
brine	*n.* 盐水
solution	*n.* 解答；解决办法；溶解；溶液
freeze	*v.* 冻结
circulate	*v.* (使)流通；(使)运行；(使)循环；(使)传播
air-cooled heat exchanger	空冷器
corrosive	*adj. & n.* 腐蚀的；腐蚀性的；腐蚀物；腐蚀剂
withstand	*vt.* 抵挡；经受住
corrosion	*n.* 侵蚀；腐蚀状态
horizontal	*adj.* 地平线的；水平的
vertical position	竖直位置
nozzle	*n.* 管口；喷嘴
draining	*n.* 排水；泄水；排泄
submerge	*v.* 浸没；淹没；掩饰
	vi. 潜水
condensate	*n.* 冷凝物
withdraw	*v.* 撤回；回收
assure	*vt.* 确保；保证；担保
fulfill	*vt.* 履行；实现；完成（计划等）
custom	*n.* 习惯；风俗
	adj. 定制的；承接定做活的
calculation	*n.* 计算；算计
automatically	*adv.* 自动地；机械地

 **Course practice**

1. Complex sentence analysis.

(1) Sometimes liquids are cooled by circulating them through pipes exposed to the atmosphere, this is called an air-cooled heat exchanger.

Sometimes liquids are cooled by circulating them through pipes exposed to the atmosphere 与 this is called an air-cooled heat exchanger 是并列句。

在前一句中：liquids 是主语；are cooled 是谓语部分；by circulating them through pipes exposed to the atmosphere 这部分作状语，其中 exposed to the atmosphere 是过去分词短语作定语修饰 pipe。

【译文】有时，液体通过暴露在空气中的管道循环冷却，这被称为空气冷却器。

(2) When a shell-and-tube heat exchanger is used for cooling, the material to be cooled may be either on the tube side or the shell side, with the cooling medium on the other side.

在句中：When a shell-and-tube heat exchanger is used for cooling 是状语从句；the material 是主语；to be cooled 是不定式作定语修饰 the material；may be either on the tube side or the shell side 是谓语部分；with the cooling medium on the other side 是介词短语作伴随状语。

【译文】当采用管壳式换热器进行冷却时，待冷却的物料可以在管侧，也可以在壳侧，冷却介质在另一侧。

(3) If a corrosive material is used on the shell side—in contact with both the inner surface of the shell and the outer surface of the tubes—both the tubes and shell must be able to withstand corrosion...

在句中：if 引导条件状语从句 a corrosive material is used on the shell side；in contact with both the inner surface of the shell and the outer surface of the tubes 是插入语，作补充说明；both the tubes and shell must be able to withstand corrosion 是主句。

【译文】如果在外壳上使用腐蚀性材料——与外壳的内表面和管的外表面都接触——管和外壳都必须耐腐蚀……

(4) If a pool of condensate builds up in such an exchanger (usually due to malfunctioning or wrongly sized steam trap) and submerges some of the tubes, the heat transfer suffers because condensing steam transfers heat more rapidly than does hot water.

在句中：If a pool of condensate builds up in such an exchanger (usually due to malfunctioning or wrongly sized steam trap) and submerges some of the tubes 是条件状语从句；the heat transfer 是主句；because condensing steam transfers heat more rapidly than does hot water 为原因状语从句。

【译文】如果冷凝水积聚在换热器内（通常是由于疏水器故障或尺寸有误造成的）并且浸没了一些管子，那么传热就会受影响，因为冷凝蒸汽比热水传热更快。

（5）The size of the shell-and-tube heat exchanger needed for any particular job depends mainly on the amount of heat to be transferred and the volume of fluid to be handled.

这是一个单句，The size of the shell-and-tube heat exchanger needed for any particular job 是主语，其中 needed for any particular job 是过去分词短语作后置定语修饰 heat exchanger。

【译文】任何特定作业所需的管壳式换热器的尺寸主要取决于要传递的热量和要处理的流体体积。

2．Translate the following paragraph.

When a liquid is used on the shellside, it is always introduced through the lower nozzle and withdrawn through the upper nozzle. In this way the engineer can be assured that the exchanger remains full, with liquid surrounding all the tubes.

Translation of the text

换热器

如果物料需要冷却，也可以使用管壳式换热器。冷却介质通常采用常温水、冷却水或冷的盐水。用盐水（盐的水溶液）是因为盐水的冻结温度比水低。有时，液体通过暴露在空气中的管道循环冷却，这被称为空气冷却器。如果用风机对管路快速吹冷风，冷却过程就会加快（图 3-4）。

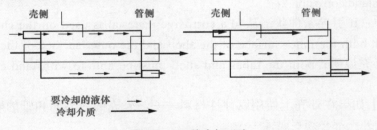

图 3-4　两种冷却形式

当采用管壳式换热器进行冷却时，待冷却的物料可以在管侧，也可以在壳侧，冷却介质在另一侧。如果两种液体中的一种腐蚀性比其他液体强（例如，盐水是有腐蚀性的），则在管侧。管体可以用耐腐蚀材料制造，而壳体用便宜的材料加工而成。如果在外壳上使用腐蚀性材料——与外壳的内表面和管的外表面都接触——管和外壳都必须耐腐蚀（图 3-5）。

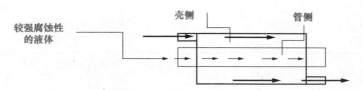

图 3-5 腐蚀性液体冷却形式

管壳式换热器可以卧式运行也可以立式运行,但常用卧式。当使用蒸汽加热时,通常从上面的孔或口进入壳体,以便从较低的口排出冷凝水。如果冷凝水积聚在换热器内(通常是由于疏水器故障或尺寸有误造成的)并且浸没了一些管子,那么传热就会受影响,因为冷凝蒸汽比热水传热更快。当液体走壳侧时,总是从下面的口进入,从上面的口引出。这样工程师才能保证换热器内充满液体,管体浸没在液体中(图 3-6、图 3-7)。

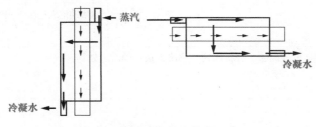

图 3-6 两种管壳式换热器示意 1

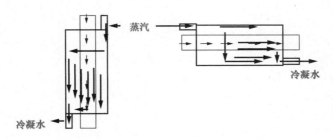

图 3-7 两种管壳式换热器示意 2

管壳式换热器应用广泛,所以在许多制造厂商的产品名单上都有。工业生产中有一些工业通用的标准设计类型。但是,没有任何一组标准的换热器能满足所有的要求,所以很多换热器是根据需要定制设计和加工的。任何特定作业所需的管壳式换热器的尺寸主要取决于要传递的热量和要处理的流体体积。大多数工程师在大学都会学习换热器的设计计算,而现在这些可以使用计算机自动完成。

Evaluate

任务名称			换热器	姓名	组别	班级	学号	日期	
考核内容及评分标准					分值	自评	组评	师评	平均分
三维目标		知识	了解换热器结构、分类及其功能等相关知识		25分				
		技能	能阅读专业英语文章、翻译专用英语词汇		40分				
		素养	在学习过程中秉承科学精神、合作精神		35分				
加分项	收获（10分）		收获（借鉴、教训、改进等）：	你进步了吗？				加分	
				你帮助他人进步了吗？					
	问题（10分）		发现问题、分析问题、解决方法、创新之处等：					加分	
总结与反思								总分	

Unit 2　Civil Aviation Maintenance

Unit objectives

【Knowledge goals】
1. 掌握空气动力学基础中涉及的举升力、质量、推力等相关知识。
2. 掌握机身结构及机身类别等相关知识。
3. 掌握地勤服务的相关知识。

【Skill goals】
1. 能够阅读专业英语文章。
2. 可以翻译专用英语词汇。

【Quality goals】
1. 在学习过程中发扬科学研究精神。
2. 增强团队合作意识。

2.1　Aerodynamics Basics

Text

民航机务

An aircraft in straight-and-level un-accelerated flight has four forces acting on it (in turning, diving, or climbing flight, additional forces come into play). These forces are: lift, an upward-acting force; drag, a retarding force of the resistance to lift and to the friction of the aircraft moving through the air; gravity, the downward effect that gravity has on the aircraft; thrust, the forward-acting force provided by the propulsion system (or, in the case of unpowered aircraft, by using gravity to translate altitude into speed). Drag and gravity are elements inherent in any object, including an aircraft. Lift and thrust are artificially created to enable an aircraft to fly (Fig. 3-8).

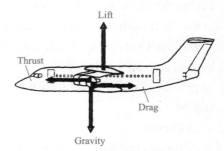

Fig. 3-8　The four forces acting on an aircraft

Lift

Bernoulli's principle states that when a fluid flowing through a tube reaches a constriction or the narrowest part of the tube, the speed of the fluid flowing through that constriction is increased and its pressure decreased. The cambered surface of an airfoil affects the airflow exactly as a constriction in a tube affects airflow.

Fig.3-9 illustrates this resemblance. The effect of air flowing over a curved surface, such as an airfoil, is similar to the effect of air flowing through a limit on the upper surface of a wing, where the air velocity is faster and the pressure is lower. In contrast, on the lower surface of the wing, the air velocity is slower and the pressure is stronger. In this way, the air creates a pressure difference (the difference in pressure) between the upper and lower surfaces of the wing, which in turn creates the lifting force of the aircraft.

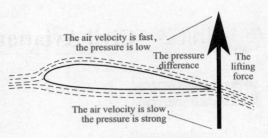

Fig. 3-9 Bernoulli's principle

It needs to be emphasized that air flowing over the top surface of the wing must reach the trailing edge of the wing in the same amount of time as the air flowing under the wing. To do this, the air passing over the top surface moves at a greater velocity than the air passing below the wing because of the greater distance it must travel along the top surface. This increased velocity, according to Bernoulli's principle, means a corresponding decrease in pressure on the surface. Thus, a pressure differential is created between the upper and lower surfaces of the wing, forcing the wing upward in the direction of the lower pressure. In other words, the difference in curvature of the upper and lower surfaces of the wing builds up the lift force.

Gravity

Gravity is a force that acts opposite to lift. Designers thus attempt to make the aircraft as light as possible. Because all aircraft designs have a tendency to increase in gravity during the development process, modern aerospace engineering staffs have specialists in the field controlling gravity from the beginning of the design. In addition, pilots must control the total gravity that an aircraft is permitted to carry (passengers, fuel, and freight) both in amount and in location. The distribution of gravity (i.e., the control of the center of gravity of the aircraft) is as important aerodynamically as the amount of gravity being carried.

Thrust

Thrust, the forward-acting force, opposed to drag as lift is opposed to gravity. Thrust is obtained by accelerating a mass of ambient air to a velocity greater than the speed of the aircraft; the equal and opposite reaction is for the aircraft to move forward. In reciprocating or turboprop-powered aircraft, thrust derives from the propulsive force caused by the rotation of the propeller, with residual thrust provided by the exhaust. In a jet engine, thrust derives from the propulsive force of the rotating blades of a turbine compressing air, which then expanded by the combustion of introduced fuel and exhausted from the engine.

Acting in continual opposition to thrust is drag, which has two primary elements, parasitic drag and induced drag.

Parasite Drag

Parasite drag includes all drags created by the airplane, except that drag directly associated with the production of lift. It's created by the disruption of the flow of air around the airplane's surfaces. Parasite drag is normally divided into three types: form drag, skin friction drag, and interference drag.

Each type of parasite drag varies with the speed of the airplane. The combined effect of all parasite drag varies proportionately to the square of the air speed. In other words, if air speed doubled, parasite drag increases by a factor of four.

Form Drag Form drag is created by any structure which protrudes into the relative wind. The amount of drag created is related to both the size and shape of the structure. For example, a square strut creates substantially more drag than a smooth or rounded strut. Streamlining reduces form drag.

Skin Friction Drag Skin friction drag is caused by the roughness of the airplane's surfaces. Even though these surfaces may appear smooth, under a microscope they may be quite rough. A thin layer of air clings to these rough surfaces and creates small eddies which contribute to drag.

Interference Drag Interference drag occurs when varied currents of air over an airplane meet and interact. This interaction creates an additional drag. One example of this type of drags is the mixing of the air where the wing and fuselage is joined.

Induced Drag

Induced drag is the main by product of the production of lift. It is directly related to the angle of attack of the wing. The greater the angle, the greater the induced drag.

Over the past several years the winglet has been developed and used to reduce induced drag. As discussed earlier in this unit, the high pressure air beneath the wing tends to spill over to the low pressure area above the wing, producing a strong secondary flow. If a winglet of the correct orientation and design is fitted to a wing tip, a rise in both total lift and drag is produced. However, with a properly designed winglet, the amount of lift produced is greater than the additional drag, resulting in a net reduction in total drag.

As the angle of attack increase, so does the drag, at a critical point, the angle of attack can become so great that the airflow is broken over the upper surface of the wing, and lift is lost while drag increases. This critical condition is termed the stall (Fig. 3-10).

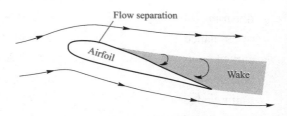

Fig. 3-10 Flow around airfoil at high angle of attack

The aerodynamics of supersonic flight is complex. The Mach number (*Ma*) refers to the method of measuring airspeed that was developed by the Austrian physicist Ernst

Mach. Mach number is the speed of an object moving through air, or any other fluid substance, divided by the speed of sound as it is in that substance for its particular physical conditions, including those of temperature and pressure.

The critical Mach number for an aircraft has been defined as that at which on some point of the aircraft, the airflow has reached the speed of sound.

At Mach numbers in excess of the critical Mach number (that is, speeds at which the airflow exceeds the speed of sound at local points on the airframe), there are significant changes in forces, pressures, and moments acting on the wing and fuselage caused by the formation of shock waves. One of the most important effects is very large increase in drag as well as a reduction in lift.

For most flights it is desirable to have all drag reduced to a minimum, and for this reason considerable attention is given to streamlining the form of the aircraft by eliminating as much drag-inducing structure as possible (e.g., retracting the landing gear, using flush riveting, and painting and polishing surfaces). Some less obvious elements of drag include the relative disposition and area of fuselage, wing, engine, and empennage surfaces; the intersection of wings and tail surfaces.

【Words and phrases】

level	*adj.* 水平的;平坦的;齐平的
un-accelerated	*adj.* 未加速的
dive	*v.* 向下俯冲
come into play	开始起作用;开始运行
retard	*v.* 延迟
resistance	*n.* 反对;抵制;抵抗;反抗;抵抗力
friction	*n.* 摩擦;摩擦力
inherent	*adj.* 固有的;内在的;与生俱来的
artificially	*adv.* 人工地;人为地
constriction	*n.* 约束;束紧
camber	*n.* 拱曲度(路面中间微拱的曲面);拱形
	v. 呈弧形;使拱起
illustrate	*v.* 图解说明;阐明
resemblance	*n.* 相似之处
curved	*adj.* 呈曲线形的
velocity	*n.* 速度
trailing edge	后缘
pressure differential	压差
curvature	*n.* 弯曲;曲率
freight	*n.* 货物
aerodynamically	*adv.* 空气动力学地

ambient	*adj.* 周围的；外界的
reciprocating	*adj.* 往复式发动机的；活塞式发动机的
propulsive	*adj.* 推进的；有推进力的
propeller	*n.* 螺旋桨；推进器
residual	*adj.* 剩余的；残留的
exhaust	*v.* 排气；耗尽
blade	*n.* 叶片；桨叶
turbine	*n.* 涡轮；涡轮机
compress	*v.* 压缩
combustion	*n.* 燃烧
parasitic	*adj.* 寄生的；附加的
induced	*adj.* 诱导的；感应的
interference	*n.* 干扰；阻碍
spill	*v.* 溢出；溅出
orientation	*n.* 方向；方位；定位；定向
winglet	*n.* （翼梢）小翼
critical	*adj.* 临界的；关键的
stall	*v. & n.* 失速
supersonic	*adj.* 超声速的
in excess of	超过
reduction	*n.* 减小；减少
eliminate	*v.* 排除；消除；消灭
retract	*v.* 收回；缩进；缩回
flush	*adj.* 齐平的
rivet	*n.* 铆钉
	vt. 铆接；固定
disposition	*n.* 配置；排列；部署
empennage	*n.* 机尾；尾翼面
intersection	*n.* 相交点

Course practice

1. Complex sentence analysis.

（1）An aircraft in straight-and-level un-accelerated flight has four forces acting on it (in turning, diving, or climbing flight, additional forces come into play).

科技英语注重对概念给予精确的定义，注意此句的主语用多个词汇的定语加以描述。

【译文】对于一架水平直飞、无加速度的飞机，有4种力作用于其上（当飞机进行转弯、俯冲、爬升飞行时，额外的力会起作用）。

（2）The cambered surface of an air-foil affects the airflow exactly as a constriction in a tube affects airflow.

① exactly as ...：正如……。

②注意 affect、effect、effort 等拼写相近的词汇词义的差别。

【译文】机翼剖面的拱形曲面对气流的影响，正如一段管子的节流处对气流的影响。

（3）As the angle of attack increases, so does the drag, at a critical point, the angle of attack can become so great that the airflow is broken over the upper surface of the wing, and lift is lost while drag increases. This critical condition is termed the stall.

注意 so do sth. 的翻译，以及 so... that 句型。

【译文】当迎角增大，阻力也随之增大，在临界点，迎角会变得非常大，以至于气流在机翼的上表面被破坏，升力会随着阻力的增加而损失。这种临界状态称为失速。

2．Translate the following sentences.

（1）Within limits, lift can be increased by increasing the angle of attack, the wing area, the free-stream velocity, or the density of the air, or by changing the shape of the airfoil.

（2）Pilots must control the total gravity that an aircraft is permitted to carry (passengers, fuel, and freight) both in amount and in location. The distribution of gravity (i.e., the control of the center of gravity of the aircraft) is as important aerodynamically as the amount of gravity being carried.

（3）Thrust is obtained by accelerating a mass of ambient air to a velocity greater than the speed of the aircraft.

（4）In reciprocating or turboprop-powered aircraft, thrust derives from the propulsive force caused by the rotation of the propeller, with residual thrust provided by the exhaust.

（5）In a jet engine, thrust derives from the propulsive force of the rotating blades of a turbine compressing air, which then expanded by the combustion of introduced fuel and exhausted from the engine.

（6）The critical Mach number for an aircraft has been defined as that at which on some point of the aircraft, the airflow has reached the speed of sound.

3．Answer the following questions.

（1）What kind of forces act on an aircraft in straight-and-level un-accelerated flight?

（2）Explain how an airfoil creates the lift for the airplane.

（3）How to obtain the thrust for the airplane?

（4）How to reduce the parasitic drag?

（5）What is Mach number? What is the critical Mach number?

（6）What will happen when airplane's speed at Mach numbers in excess of the critical Mach number?

4. Fill in the following blanks according to the text.

(1) An aircraft in straight-and-level un-accelerated flight has four forces acting on it, they are lift, _____; drag, _____; gravity, _____; and thrust, _____.

(2) Acting in continual opposition to thrust is drag, which has two primary elements: _____ drag and _____ drag.

(3) Parasite drag is normally divided into three types: _____, _____, _____.

Translation of the text

空气动力学基础

对于一架水平直飞、无加速度的飞机，有4种力作用于其上（当飞机进行转弯、俯冲、爬升飞行时，额外的力会起作用）。这些力是：升力，一种向上作用的力；阻力，对升力的阻力和飞机在空气中移动时的摩擦力；重力，重力对飞机产生向下作用的力；推力，由推进系统提供的向前作用的力（或者在无动力的情况下，通过重力将高度转化为速度）。阻力和重力是任何物体固有的，包括飞机；升力和推力是人为制造的，使飞机能够飞行（图3-8）。

升力

伯努利原理指出，当流体流过管道到达收缩处或管道最窄的部分时，速度提高，压力降低。机翼剖面的拱形曲面对气流的影响，正如一段管子的节流处对气流的影响。

图3-9说明了伯努利原理。空气流过一个弯曲的表面，如翼型，其效果类似于空气通过一个限制，在机翼上表面，空气流速较快，压强小。与之相反，在机翼下表面，空气流速较慢，压强大。这样，空气在机翼上下表面形成了压力差（压强差），进而形成了飞机的升力。

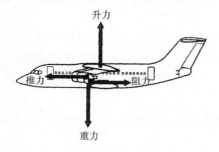

图3-8 作用在飞机上的四个力

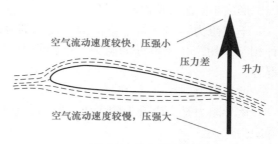

图3-9 伯努利原理

需要强调的是，流过机翼上表面的空气必须与流过机翼下表面的空气在相同的时间内到达机翼后缘。要做到这一点，通过顶部表面的空气要比通过机翼下方的空气以更大的速度移动，因为它必须沿着顶部表面移动更长的距离。根据伯努利原理，速度的增加意味着表面压力的相对降低。因此，在机翼的上下表面之间产生压力差，迫使机翼沿着较低压力的方向向上。换句话说，机翼上下表面的曲率差异形成了升力。

重力

重力是一种与升力方向相反的力。因此,设计师试图使飞机尽可能地轻。由于所有的飞机在设计过程中都有增加重力的趋势,现代航空航天工程从设计开始就有控制重力的专家。此外,飞行员必须控制飞机允许携带的总重力(乘客、燃料和货物)、数量和位置。重力的分布(即飞机重心的控制)在空气动力学上和承载的重力一样重要。

推力

推力,是向前作用的力,与阻力方向相反,就像升力与重力方向相反一样。推力是通过将周围空气加速到比飞机速度更快的速度而获得的;相等和相反的作用力使飞机向前移动。在往复式或涡轮螺旋桨飞机中,推力来自螺旋桨旋转产生的推进力,剩余推力由排气提供。在喷气发动机中,推力来自涡轮机旋转叶片压缩空气的推进力,然后通过引入燃料的燃烧而膨胀,并从发动机中排出。

与推力持续相反的是阻力,它有两个要素:寄生阻力和诱导阻力。

寄生阻力

寄生阻力包括飞机产生的所有阻力,除了与产生升力直接相关的阻力。它是由飞机表面周围空气流动的中断造成的。寄生阻力通常分为三种类型:形式阻力、表面摩擦阻力和干扰阻力。

每种寄生阻力的大小随飞机的速度而变化。所有寄生阻力的综合效应与空气速度的平方成正比。换句话说,如果空气速度增加一倍,寄生阻力会增大四倍。

形式阻力 形式阻力是由在相对风中任何凸出的结构产生的。产生的阻力量与结构的大小和形状有关。例如,方形支柱产生的阻力比光滑或圆形支柱大得多,流线型减少了形式阻力。

表面摩擦阻力 表面摩擦阻力是由飞机表面的粗糙度引起的。尽管这些表面看起来很光滑,但在显微镜下它们相当粗糙。一层薄薄的空气附着在这些粗糙的表面上,形成小涡流,从而产生阻力。

干扰阻力 当飞机上不同的气流相遇并相互作用时,就会产生干扰阻力,这种相互作用产生了额外的阻力。这种阻力的一个例子是机翼和机身连接处的空气混合。

诱导阻力

诱导阻力是产生升力的主要副产物,它与机翼的迎角直接相关,角度越大,诱导阻力越大。

在过去的几年里,小翼已经被开发出来并用于减小诱导阻力。正如本单元前面所讨论的,机翼下方的高压空气倾向于溢出到机翼上方的低压区域,产生强烈的二次流。如果在翼尖安装方向和设计都正确的小翼,就会导致总升力和总阻力增加。如果小翼设计得当,产生的升力会大于额外的阻力,最终减小总阻力。

当迎角增大,阻力也随之增大,在临界点,迎角会变得非常大,以至于气流在机翼的上表面被破坏,升力会随着阻力的增加而损失。这种临界状态称为失速(图3-10)。

超声速飞行的空气动力学是复杂的。马赫数(Ma)是指由奥地利物理学家恩斯特·马赫发明的测量空速的方法。马赫数是物体穿过空气或任何其他流体物质的速度,除以该物质在特定物理条件(包括温度和压力)下的声速。

飞机的临界马赫数被定义为在飞机的某一点上,气流达到声速时的马赫数。

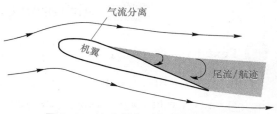

图 3-10　在大迎角时气流围绕翼型流动

当马赫数超过临界马赫数(即在机体局部点气流速度超过声速)时,由于激波的形成,作用在机翼和机身上的压力和力矩会发生显著变化。最重要的影响之一是阻力的大幅增大以及升力的减小。

对于大多数飞行来说,将所有的阻力降低到最小是可取的,因此,尽可能多地消除拖曳结构(如收起起落架、使用平头铆接以及油漆和抛光表面)以简化飞机的形式。一些不太明显的阻力因素包括机身、机翼、发动机和尾翼表面的相对布局和面积,机翼和尾部表面的交点。

Evaluate

任务名称		空气动力学基础	姓名	组别	班级	学号	日期
考核内容及评分标准			分值	自评	组评	师评	平均分
三维目标	知识	空气动力学基础中涉及的升力、重力、推力等相关知识	25 分				
	技能	能阅读专业英语文章、翻译专用英语词汇	40 分				
	素养	在学习过程中秉承科学精神、合作精神	35 分				
加分项	收获 (10 分)	收获(借鉴、教训、改进等):	你进步了吗?			加分	
			你帮助他人进步了吗?				
	问题 (10 分)	发现问题、分析问题、解决方法、创新之处等:				加分	
总结与反思						总分	

2.2 Airframe Construction

 Text

The airframe of a fixed-wing aircraft is generally considered to consist of five principal units: the fuselage, wing, stabilizers, flight control surface, and landing gear (powerplant of the airplane is normally optional) (Fig. 3-11, Fig. 3-12). Helicopter airframes consist of the fuselage, main rotor (on helicopters with a single main rotor) and related gearbox, tail rotor, and the landing gear.

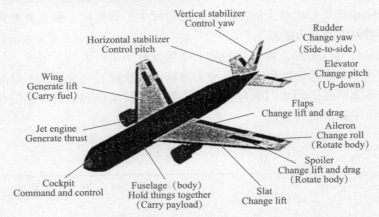

Fig. 3-11 The main parts of an airplane

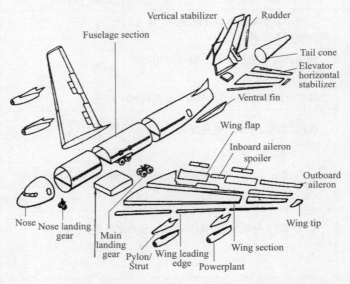

Fig. 3-12 Typical structural components of a turbine powered aircraft

The airframe components are constructed from a wide variety of materials and are joined by riveting, bolting, screwing, welding or adhesive. The aircraft components are composed of various parts called structural members (i.e., stringers, longerons, ribs, bulkheads, etc.).

Fuselage

The fuselage is the main structure or body of the aircraft. It provides space for cargo, controls, accessories, passengers, and other equipment. In single engine aircraft, it also houses the powerplant. In multi-engine aircraft, the engines may either be in the fuselage, attached to the fuselage, or suspended from the wing structure. They vary principally in size and arrangement of the different compartments.

The typical design of fuselage may be divided into three classes: ① monocoque, ② semi-monocoque, ③ reinforced shell. The true monocoque construction (Fig. 3-13) uses formers, frame assemblies, and bulkheads to give shape to the fuselage, but the skin carries the primary stresses. Since no bracing members are present, the skin must be strong enough to keep the fuselage rigid. Thus, the biggest problem involved in monocoque construction is maintaining enough strength while keeping the mass within allowable limits.

To overcome the strength/mass problem of monocoque construction, a modification called semi-monocoque construction (Fig. 3-14) was developed. In addition to formers, frame assemblies, and bulkheads, the semi-monocoque construction has the skin reinforced by longitudinal members.

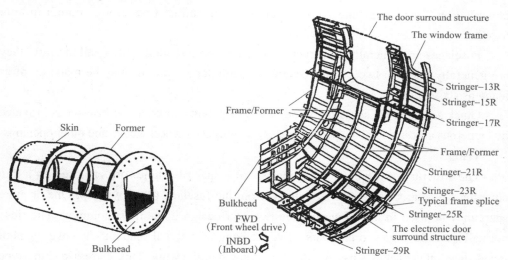

Fig. 3-13 Monocoque construction　　　　Fig. 3-14 Semi-monocoque construction

The reinforced shell has the skin reinforced by a complete framework of structural members.

Different portions of the same fuselage may belong to any one of the three classes, but most aircraft are considered to be of semi-monocoque type construction.

The semi-monocoque fuselage is constructed primarily of the alloys of aluminum magnesium, although steel and titanium are found in areas of high temperatures.

Primary bending loads are taken by the longerons, which usually extend across several points of support. The longerons are supplemented by other longitudinal members, called stringers. Stringers are more numerous and lighter in mass than longerons. The stringers are served as fillings. They have some rigidity, but are chiefly used for giving shape and for attachment of the skin. Stringers and longerons prevent tension and compression from bending the fuselage.

The vertical structural members are referred to as bulkheads, frames, and formers. The heaviest of these vertical members are located at intervals to carry concentrated loads and at points where fittings are used to attach other units (such as the wings, powerplants, and stabilizers).

The strong, heavy longerons hold the bulkheads and formers, and these, in turn, hold the stringers. All of these joined together form a rigid fuselage framework.

The metal skin or covering is riveted to the longerons, bulkheads, and other structural members and carries part of the load.

There are a number of advantages in the use of the semi-monocoque fuselage. The bulkheads, frames, stringers, and longerons facilitate the design and construction of a streamlined fuselage, and add to the strength and rigidity of the structure. The main advantage, however, lies in the fact that it does not depend on a few members for strength and rigidity. This means that a semi-monocoque fuselage, because of its stressed-skin construction, may withstand considerable damage and still be strong enough to hold together.

Fuselages are generally constructed in two or more sections. On small aircraft, they are generally made in two or three sections, while larger aircraft may be made up of as many as six sections.

Quick access to the accessories and other equipment carried in the fuselage is provided by numerous access doors, inspection plates, landing gear wheel wells, and other openings.

Wing

The structure of a typical wing consists of a spar-rib framework enclosed by a thin covering of metal sheet. The spar extends from the fuselage to the wingtip. One or more spars may be used in the wing, but the two-spar design is the most common. The ribs, normally at right angles to the spars, give the wing its external shape. If the covering is of metal sheet, it contributes its own share of strength to the wing. This "stressed skin" type of wing is used in all large planes.

The size and shape of wing vary widely, depending on specific aerodynamic considerations. Wings of many supersonic planes have a high degree of sweepback (arrowhead tapering from the nose of the plane) and are as thin as possible, with a knife-like leading edge. Such a shape helps to reduce the shock of compression when the plane approaches the speed of sound.

【Words and phrases】

airframe	n. 机体；飞机骨架
fixed-wing	n. 固定翼

fuselage	n. 机身
stabilizer	n. 稳定面
control surface	操纵面（舵面）
landing gear	起落架
helicopter	n. 直升机
rudder	n. 方向舵
elevator	n. 升降舵
flap	n. 襟翼
aileron	n. 副翼
spoiler	n. 扰流板
slat	n. 缝翼
component	n. 元件；部件
structural member	结构件
stringer	n. 纵梁；桁条；长桁
longeron	n. 纵梁；大梁
bulkhead	n. 隔框（通常指封闭的或半封闭的隔框）
tail cone	尾锥
wing tip	翼尖
pylon	n. 桥塔；指示塔
strut	挂架
main landing gear	主起落架
nose landing gear	前起落架
cargo	n. 货物
monocoque	n. 单体壳式（结构）
semi-monocoque	半单体式
frames/former	构架；构筑物
assembly	n. 组装；组件
strength	n. 力量
framework	n. 框架；边框；隔框（通常指空心隔框）
portion	n. 部分
rigidity	n. 刚度
streamlined	adj. 流线型的
stress	n. 压力；应力；加压力
withstand	v. 经受；承受
access door	检修门
contribute to...	有助于……
consideration	n. 考虑；需要考虑的事项
sweepback	n. 后掠角
leading edge	前部；前缘

Course practice

1. Complex sentence analysis.

（1）The airframe components are constructed from a wide variety of materials and are joined by riveting, bolting, screwing, welding or adhesive.

联系前后文，这句中的"riveting, bolting, screwing"应翻译成"铆接、螺栓连接、螺钉连接"。

【译文】机体部件由各种材料制成，并通过铆接、螺栓连接、螺钉连接、焊接或黏合剂连接。

（2）The metal skin or covering is riveted to the longerons, bulkheads, and other structural members and carries part of the load.

be riveted to 表示被铆接在……上。

【译文】金属蒙皮或覆盖物被铆接在大梁、舱壁和其他结构构件上，并承担部分载荷。

2. Translate the following paragraphs.

（1）The aircraft components are composed of various parts called structural members (i.e., stringers, longerons, ribs, bulkheads, etc.).

（2）The fuselage is the main structure or body of the aircraft. It provides space for cargo, controls, accessories, passengers, and other equipment.

（3）In single engine aircraft, it also houses the powerplant. In multi-engine aircraft, the engines may either be in the fuselage, attached to the fuselage, or suspended from the wing structure.

3. Fill in the following blanks according to the text.

（1）The airframe of a fixed-wing aircraft is generally considered to consist of five principal units: _____, _____, _____, _____, and _____. Helicopter airframes consist of _____, _____ (on helicopters with a single main rotor) and _____, _____, and _____.

（2）The airframe components are constructed from a wide variety of materials and are joined by _____, _____, _____, _____ or _____.

（3）The typical design of fuselage may be divided into three classes: ① _____, ② _____, or ③ _____.

Translation of the text

机体构造

固定翼飞机的机体通常被认为由机身、机翼、尾翼、飞控面和起落架五个主要部件组成（飞机的动力装置通常是可选的）（图3-11、图3-12）。直升机机体由机身、主旋翼（在直升机上只有一个主旋翼）和相关齿轮箱、尾旋翼和起落架组成。

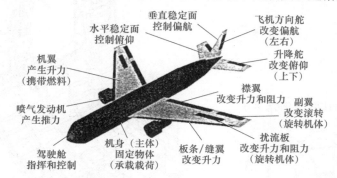

图 3-11　飞机主要部件示意

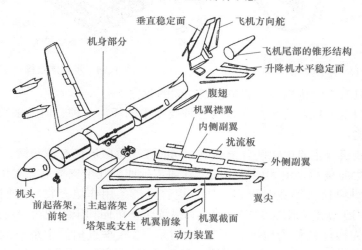

图 3-12　涡轮飞机典型结构件

机体部件由各种材料制成，并通过铆接、螺栓连接、螺钉连接、焊接或黏合剂连接。飞机部件由称为结构部件的各种部件（即纵梁、大梁、肋、舱壁等）组成。

机身

机身是飞机的主要结构或主体，它为货物、控制装置、附件、乘客和其他设备提供空间。在单引擎飞机上，它也有动力装置。在多引擎飞机中，发动机可以安装在机身内，也可以安装在机身上，或者悬挂在机翼结构上。它们主要在大小和在不同舱室的排列上有所不同。

典型的机身设计可分为三类：①单体式；②半单体式；③加固壳式。真正的单体结构（图3-13）使用成型器、框架组件和舱壁来给机身塑形，但外壳承担主要的应力。由于没有支撑构件，所以蒙皮必须足够坚固以保持机身的刚性。因此，单体结构所涉及的最大问题是保持足够的强度，同时将质量保持在允许的范围内。

为了克服单体结构的强度、质量问题，一种被称为半单体的改进结构（图3-14）被开发出来。除构架、框架组件和舱壁外，半单体结构还具有纵向构件加固的蒙皮。

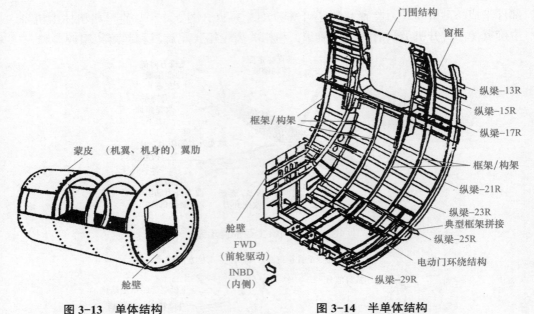

图 3-13 单体结构　　　　图 3-14 半单体结构

加固壳具有由结构构件的完整框架加固的蒙皮。

同一机身的不同部分可能属于这三种类型中的任何一种，但大多数飞机被认为是半单体结构。

半单体式机身主要由铝镁合金制成，同时在高温区域也有钢和钛。

主要的弯曲载荷是由大梁承担的，它通常延伸到几个支撑点。大梁由其他纵向构件补充。纵梁比大梁数量更多、质量更轻。桁条用作填充物。它们有一定的刚性，但主要用于塑造和附着蒙皮。大梁和纵梁防止拉伸和压缩使机身弯曲。

垂直结构构件被称为舱壁、框架和构架。这些垂直构件中最重的部分位于承载集中载荷的间隔处，以及用于连接其他部件（如机翼、动力装置和稳定面）的连接处。

结实而沉重的大梁支撑着舱壁和构架，而舱壁和构架又支撑着纵梁。所有这些连接在一起形成一个坚固的机身框架。

金属蒙皮或覆盖物被铆接在大梁、舱壁和其他结构构件上，并承担部分载荷。

使用半单体式机身有许多优点。舱壁、框架、纵梁和大梁便于流线型机身的设计和构造，并增加了结构的强度和刚性。然而，它的主要优点在于不依赖少数构件的强度和刚性。这意味着半单体机身，由于它的应力蒙皮结构可以承受相当大的损伤，但仍然足够坚固地凝结在一起。

机身通常分为两个或更多部分。在小型飞机上，机身通常由 2~3 个部分组成，而大型飞机可能由多达 6 个部分组成。

通过许多检修门、检查板、起落架轮舱和其他开口，可以快速进入机身内检修附件和其他设备。

机翼

典型的机翼结构由一层薄薄的金属板覆盖的翼肋框架组成。翼梁从机身延伸到翼尖。机翼可以使用一个或多个梁，但双梁设计是最常见的。翼肋通常与翼梁成直角，形成机翼的外部形状。如果覆盖物是金属板，它也能为机翼提供一定的强度。这种"受力蒙皮"型机翼在所有大型飞机上都有应用。

机翼的大小和形状差异很大，取决于具体的空气动力学因素。许多超声速飞机的机翼都有高度的后掠角（箭头从飞机机头逐渐变细），并且尽可能薄，有一个刀状的前缘。当飞机接近声速时，这种形状有助于减少压缩带来的冲击。

Evaluate

任务名称		机体构造		姓名	组别	班级	学号	日期	
考核内容及评分标准				分值	自评	组评	师评	平均分	
三维目标	知识	了解机身结构及机身类别等相关知识		25分					
	技能	能阅读专业英语文章、翻译专用英语词汇		40分					
	素养	在学习过程中秉承科学精神、合作精神		35分					
加分项	收获（10分）	收获（借鉴、教训、改进等）：			你进步了吗？			加分	
					你帮助他人进步了吗？				
	问题（10分）	发现问题、分析问题、解决方法、创新之处等：						加分	
总结与反思								总分	

2.3 Basic Knowledge of Ground Handling and Servicing

Text

The common servicing tasks for an aircraft in line maintenance include parking, towing, wheel chocking, fitting and removal of the protection devices of an aircraft, connecting ground power and pneumatic supplies, water servicing (fill and drain), fuelling and de-fuelling, air and oxygen charging, necessary cleaning, de-icing arid removal of snow and frost and so on.

The following is some basic knowledge about ground handling and servicing in detail.

Jacking the Aircraft

Aircraft must often be raised from a hangar floor for weighing, maintenance or repair.

There are several methods for doing this. You should follow the aircraft manufacturer's instructions. If an aircraft slips out of a hoist or falls off a jack (Fig. 3-15), the cost to repair the aircraft is usually quite high.

Many aircraft have stressed panels that must be installed before an aircraft is jacked, hoisted, or moved. Most modern aircraft have jack pads located on their main wing spars (Fig. 3-16).

It is often necessary to lift only one wheel from the floor to change a tire or to service a wheel or brake. For this type of jacking, some manufacturers have made provisions on the strut for the placement of a short hydraulic jack (Fig. 3-17). When using this method, never place the jack under the brake housing or in any location that is not specifically approved by the manufacturer.

Fig. 3-15　Typical tripod jack　　Fig. 3-16　Forward jack pad　　Fig. 3-17　Short hydraulic jack

It is usually recommended that both wheels not be lifted off the floor at the same time when jacking from the landing gear struts.

When jacked from the struts, some aircraft have a tendency to move sideways and tilt the jack as the mass is removed from the tire. If this should occur, lower the jack and straighten it, and then raise the wheel again. To keep the aircraft from moving while it is on the jack, the wheels that are not jacked should be securely chocked. Fig. 3-18 and Fig. 3-19 show some typical kinds of chocks.

 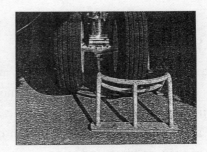

Fig. 3-18　Rudder wheel chock　　Fig. 3-19　Metallic wheel chock

The most important consideration when jacking an aircraft is to follow the manufacturer's instructions in detail. Be sure to use the proper jacks so that the aircraft remains level with no tendency for it to slip off the jacks. Most higher capacity jacks have screw-type safety collars to prevent the jack from inadvertently retracting. Be sure that these collars are screwed down as the airplane is raised. Jacks that do not have the screw-type safety usually have holes drilled in the shaft so lock pins can be inserted to guard against the jack retracting.

Guard against any movement within the aircraft when it is on jacks, since shifting the mass behind the jack could cause the aircraft to tilt enough to fall off the jack.

Before lowering the aircraft, be sure to remove workstands, ladders and other equipment. Items placed under the aircraft while it was on jacks could cause damage when the aircraft is lowered. Furthermore, be sure that the landing gear is down and locked before the aircraft is lowered evenly.

Taxiing Aircraft

Airplanes and helicopters are designed to fly, and movement on the ground is often a rather awkward procedure. Because of this, only qualified persons authorized to taxi aircraft may actually taxi an aircraft.

Before starting an engine, be sure that the areas in front and behind the aircraft are clear of people and equipment. A maintenance technician should be checked out by a properly qualified instructor before taxiing a new or different aircraft.

From the cockpit, it is difficult to assure that there is sufficient clearance between the aircraft structure and any buildings or other aircraft. Therefore, it is a good policy to station signalmen where they can watch the wings or rotor and any obstructions. When this is done, it is important that all personnel use the same signals and understand exactly what the signals mean to avoid misunderstanding at a crucial time (Fig. 3-20). The signalman has the responsibility of remaining in a position that is visible from the cockpit. To ensure that signalman can be seen at all times, make sure that signalmam can see the pilot's eyes while directing him.

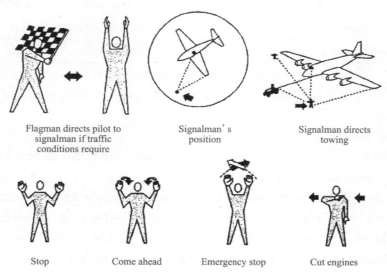

Fig. 3-20 Standard hand signals allow ground personnel to direct the movement of aircraft

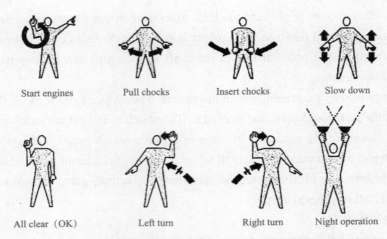

Fig. 3-20 Standard hand signals allow ground personnel to direct the movement of aircraft (continued)

When taxiing an aircraft at a tower-controlled airport, typically the signalman must receive a clearance from ground control before taxiing. Located in the cab of the control power is a powerful light that controllers can use to direct light beams of various colors toward aircraft. Each color or color combination has a specific meaning for an aircraft on the airport surface (Table 3-1).

Table 3-1 Color and type of signal

Color and type of signal	Meaning on the ground
Steady green	Cleared for takeoff
Flashing green	Cleaned to taxi
Steady red	Stop
Flashing red	Taxi clear of landing area(runway) in use
Flashing white	Return to starting point on airport
Alternating red and green	Exercise extreme caution

Once the aircraft is in motion, immediately tap the brakes to insure they are working properly. After testing the brakes, test the nose landing gear steering system to make sure it is operating.

Towing Aircraft

It is often necessary to move an aircraft without using its engines. This can be accomplished by towing the aircraft.

Large aircraft are towed with a tractor (special towing vehicle), connected by a special tow bar (Fig. 3-21), or towed by a towbarless aircraft towing tractor (Fig. 3-22). Extreme care must be used to avoid towing an aircraft too fast and to be sure that here is always sufficient clearance between the wings and any obstructions.

Fig. 3-21　Tow bar and towing tractor　　　Fig. 3-22　Towbarless aircraft towing tractor

When an aircraft is being towed, a qualified person should be in the cockpit to operate the aircraft brakes when needed since the brakes on a towing vehicle are usually insufficient to overcome a large aircraft's momentum. Extra personnel should be assigned to watch the wingtips and tail for clearance between other objects.

The nose landing gear on most aircraft has a very definite limit to the amount that can be tamed. When towing, it is easy to exceed this limit. If the turning radius is exceeded, the nose landing gear strut and steering mechanism will be damaged. Damage can be quite extensive, requiring replacement of the nose landing gear shock strut. Some aircraft have a method of disconnecting a locking device so the nose wheel can be swiveled to facilitate maneuvering. If this is the case, the locking device must always be disconnected when an aircraft is towed. Furthermore, remember to reset the lock after removing the tow bar from the aircraft. Persons riding in the aircraft should not attempt to steer the nose wheel when a tow bar is attached to the aircraft.

Foreign Object Damage(FOD)

If foreign objects such as nuts, bolts, and safety wire are drawn into the inlet of a turbine engine, or through the arc of a rotating propeller blade, they can easily cause damage that can lead to catastrophic failure. Furthermore, if these objects are caught in an engine's high-velocity exhaust and strike another object, these objects can cause serious injury. Therefore, it is extremely important that an airport flight line be kept clean. Furthermore, you, as an aircraft maintenance technician, should develop the habit of picking up all loose hardware and rags you find on the ramp and deposit them in a suitable container.

Your personal safety is much more important than damage to equipment. Therefore, be aware that propellers and jet intakes make aircraft operation areas extremely dangerous places. Sound maintenance practice requires that aircraft operating their engines must illuminate their navigation lights and rotating beacons. Furthermore, a second person should be stationed on the ground to warn anyone in the area. If it is necessary to approach an aircraft whose engine is running, be sure that you have the attention of the person in the cockpit so they know where you are at all times. The danger of a propeller is so obvious that it hardly needs mention, but sadly enough, many people have been killed by walking into a spinning propeller. Always be cautious of rotating propellers.

Turbojet aircraft are dangerous from both the front and the rear. When an engine is running, it moves a large volume of air and produces a low pressure area in front of the intake (Fig. 3-23). This low pressure area can draw a person into the engine, and can certainly ingest such items as hats, clipboards, and loose items of clothing. Behind the aircraft in the exhaust area, the high-velocity exhaust can cause severe damage to both people and equipment.

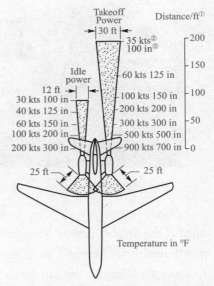

Fig. 3-23 Stay clear of the dangerous areas

Safety Around Compressed Gases

Compressed gases are found in all aircraft maintenance shops. For example, compressed air powers pneumatic drill motors, rivet guns, paint spray guns, and cleaning guns. In addition, compressed nitrogen is used to inflate tires and shock struts while compressed acetylene is used in welding.

Most shop compressed air is held in storage tanks and routed throughout the shop in high pressure lines. This high pressure presents a serious threat of injury. For example, if a concentrated stream of compressed air is blown across a cut in the skin, it is possible for the air to enter the bloodstream and cause severe injury or death. Be very careful when using compressed air and not to blow dirt or chips into the face of anyone standing nearby.

To prevent eye injury, you should wear eye protection when using pneumatic tools. To prevent injury from a ruptured hose, always keep air hoses and fittings in good condition.

Far too many accidents occur when inflating or deflating tires. Always use calibrated tire gauges, and make certain to use a regulator that is in good working condition.

High pressure compressed gases are especially dangerous if they are mishandled. Oxygen and nitrogen are often found in aviation maintenance shops stored in steel cylinders under a pressure of around 2,000 psi (1 psi=6.895 kPa). These cylinders have brass valves screwed into them. If a cylinder is knocked over and the valve broken off, the escaping high pressure gas would propel the tank like a rocket. Because this would create a substantial hazard, you should make sure that all gas cylinders are properly supported. A common method of securing high pressure cylinders in storage is by chaining them to a building. Furthermore, a cap should be securely installed on any tank that is not connected

---

① 1 ft=0.304 8 m.
② 1 kts=1.852 km/h.
③ 1 in=2.54 cm.

into a system. This protects the valve from damage.

It is extremely important that oxygen cylinders be treated with special care. In addition to having all of the dangers inherent with other high pressure gases, oxygen always possesses the risk of explosion. For example, you must never allow oxygen to come in contact with petroleum products such as oil or grease, since oxygen causes these materials to ignite spontaneously and burn. Furthermore, never use an oily rag, or tools that are oily or greasy, to install a fitting or a regulator on an oxygen cylinder.

【Words and phrases】

tow	v. & n. 拖（飞机）
chock	n. 轮挡
removal	n. 拆除
fuelling/defueling	n. 加油/抽油
oxygen charging	氧气灌充
de-ice/ anti-ice	v. 除冰/防冰
frost	v. 结霜
	n. 霜
jacking	n. 顶升
hangar	n. 机库
hoist	v.（用起重机）吊起
	n. 起重器械
jack	n. 千斤顶
provision	n. 提供
taxiing	v. 滑行
clearance	n. 净空；许可
signalman	n. 信号员
obstruction	n. 障碍；阻碍
personnel	n. 人员；职员
inlet/outlet	n. 入口/出口
catastrophic	adj. 灾难性的；惨重的
injury	n. 伤害；损害
rag	n. 碎屑；破布
ramp	n. 停机坪；坡道；活动舷梯
deposit	n. 放置；存放；沉淀
be aware that...	意识到……
illuminate	v. 照明；点亮
beacon	n. 信标
approach	v. 接近
spin	v. 旋转

ingest	v. 吞下；获取
nitrogen	n. 氮气
acetylene	n. 乙炔
inflate/deflate	v.（轮胎）充气/放气
brass	n. 黄铜
hazard	n. 危险
	v. 冒险
petroleum	n. 石油；石油制品
grease	n. 油脂
spontaneously	adv. 自发地
greasy	adj. 涂有油脂的；油腻的

Course practice

1. Complex sentence analysis.

(1) For this type of jacking, some manufacturers have made provisions on the strut for the placement of a short hydraulic jack. When using this method, never place the jack under the brake housing or in any location that is not specifically approved by the manufacturer.

注意此句中 never do sth. 和 any sth. that is not 表达的是强调否定的意思。

【译文】对于这种类型的顶升，有的制造商在支柱上预留了可放置这种短液压千斤顶的位置。当使用此法顶升时，千万不要将千斤顶顶在制动器壳体下或任何未经厂家许可的位置。

(2) If this should occur, lower the jack and straighten it, and then raise the wheel again.

这里的 should 不是应该的意思，是典型的虚拟语气句型，应翻译为如果，万一。

【译文】如果这种情况发生了，应放低千斤顶并扶正，然后重新顶升机轮。

(3) ...only qualified persons authorized to taxi aircraft may actually taxi an aircraft.

此句为强调语句，过去分词短语 authorized to taxi aircraft 作定语，修饰 qualified persons。

【译文】只有被授权（指挥操作）飞机滑行的合格人员才能实际（指挥）飞机滑行。

2. Translate the following paragraphs.

(1) If an aircraft slips out of a hoist or falls off a jack, the cost to repair the aircraft is usually quite high.

(2) It is usually recommended that both wheels not be lifted off the floor at the same time when jacking from the landing gear struts.

(3) Airplanes and helicopters are designed to fly, and movement on the ground is

often a rather awkward procedure.

（4）From the cockpit, it is difficult to assure that there is sufficient clearance between the aircraft structure and any buildings or other aircraft. Therefore, it is a good policy to station signalmen where they can watch the wings or rotor and any obstructions.

（5）If foreign objects such as nuts, bolts, and safety wire are drawn into the inlet of a turbine engine, or through the arc of a rotating propeller blade, they can easily cause damage that can lead to catastrophic failure.

（6）Far too many accidents occur when inflating or deflating tires. Always use calibrated tire gauges, and make certain to use a regulator that is in good working condition.

3. Fill in the following blanks and table according to the text.

（1）Fill in the following table according to Table 3-1.

Color and type of signal	Meaning on the ground
Steady green	
Flashing green	
Steady red	
Flashing red	
Flashing white	
Alternating red and green	

（2）Turbojet aircraft are dangerous from both the front and the rear. According to Fig.3-23, _____ ft is minimum safety distance from the front, _____ ft and _____ ft minimum distance for idle power and takeoff power from the rear.

Translation of the text

地勤服务基本知识

飞机在航线维护中的常见维修任务包括停放、牵引、轮塞、安装和拆卸飞机的保护装置、连接地面电源和气动供应、供水（加水和排水）、加油和抽油、充气和充氧、必要的清洁、除冰、除雪和除霜等。

以下是一些有关地勤服务的基本知识。

顶升飞机

为了称重、维护或修理，必须经常从机库的地板上抬起飞机。

有几种方法可以做到这一点，应该按照飞机制造商的说明去做。如果一架飞机滑出起重机或从千斤顶（图3-15）上掉下来，修理飞机的费用通常是相当高的。

许多飞机都有应力板，必须在飞机被顶起、吊起或移动之前安装。大多数现代飞机的主翼梁上有千斤顶垫（图3-16）。

换轮胎或维修车轮和制动器时，通常只需从地板上抬起一个轮子。对于这种类型的顶升，有的制造商在支柱上预留了可放置这种短液压千斤顶（图3-17）的位置。当使用此法顶升时，千万不要将千斤顶顶在制动器壳体下或任何未经厂家许可的位置。

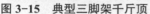

图 3-15　典型三脚架千斤顶　　　图 3-16　正顶垫　　　图 3-17　短液压千斤顶

通常建议，当顶起起落架支柱时，两个轮子不要同时离开地面。

当从支柱上取下千斤顶时，一些飞机有向侧面移动的倾向，千斤顶随着质量从轮胎上卸下而倾斜。如果这种情况发生了，应放低千斤顶并扶正，然后重新顶升机轮。为了防止飞机在千斤顶上移动，没有被千斤顶顶起的轮子应该被牢牢地卡住。图 3-18 和图 3-19 所示为典型的固定类型。

图 3-18　舵轮座　　　　　　　图 3-19　金属挡轮

顶升飞机时最重要的考虑因素是要严格按照制造商的说明操作。一定要使用适当的千斤顶，使飞机保持水平，不会从千斤顶上滑落。大多数大负载的千斤顶都有螺杆式安全环，以防千斤顶无意中缩回。飞机升起时，一定要把这些顶圈拧紧。没有螺栓式安全装置的千斤顶通常在轴上有钻孔，这样可以插入锁销来防止千斤顶缩回。

当飞机在千斤顶上时，要防止飞机内部的任何运动，因为在顶升后的质量移动可能会使飞机倾斜以致从千斤顶上掉下来。

在下放飞机之前，请务必拆除工作台、梯子和其他设备。当飞机在千斤顶上时，放在飞机下面的物品可能会在飞机下降时造成损坏。此外，在飞机均匀下降之前，要确保已放下起落架并将其锁定。

滑行的飞机

飞机和直升机是为飞行而设计的，在地面上移动往往是相当不便的过程。因此,只有被授权(指挥操作)飞机滑行的合格人员才能实际(指挥)飞机滑行。

在起动发动机之前,请确保飞机前后区域没有人员和设备。在滑行一架新的或不同类型的飞机之前,应该由具备适当资质的维修技术员进行检查。

从驾驶舱中,很难保证飞机结构与任何建筑物或其他飞机之间有足够的间隙。因此,最好将信号员安置在可以观察机翼或旋翼和任何障碍物的地方。此时,重要的是所有人员使用相同的信号,并准确理解信号的含义(图3-20),以免在关键时刻产生误解。信号员有责任留在从驾驶舱内可见的位置。为了确保信号员在任何时候都能被看到,请保证信号员在指挥飞行员的时候能看到飞行员的眼睛。

当飞机在塔台控制的机场滑行时,信号员通常必须在开始滑行前获得地面控制中心的

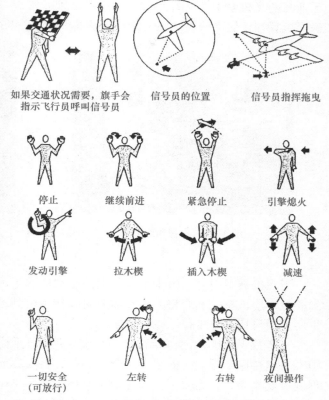

图 3-20　标准手势使地面人员能够指挥飞机的运动

许可。位于驾驶室的控制电源是强光,控制器可以使用各种颜色的光束引导飞机。机场地面上每一种颜色或颜色组合对一架飞机都有特定的含义(表3-1)。

表 3-1　信号的颜色和类型

信号的颜色和类型	现场含义
绿灯常亮	清空跑道便于起飞
绿灯闪烁	清空跑道便于滑行
红灯常亮	停靠/停止
红灯闪烁	滑出正在使用的着陆区(跑道)
白灯闪烁	返回机场起点
红绿灯交替闪烁	格外小心

一旦飞机开始飞行,立即踩下制动踏板以确保它们正常工作。测试制动后,测试前起落架转向系统以确保其正常工作。

牵引飞机

在不使用发动机的情况下移动飞机常常是必要的。这可以通过牵引飞机来完成。

大型飞机用牵引车（专用牵引车辆）牵引，用专用拖杆连接（图3-21），或者用无拖杆的飞机牵引车（图3-22）牵引。必须非常小心，以避免牵引飞机时太快，并确保机翼和任何障碍物之间始终有足够的间隙。

图3-21　拖杆和牵引车

图3-22　无拖杆的飞机牵引车

当飞机被牵引时，由于牵引车上的制动通常不足以克服大型飞机的动量，因此驾驶舱中应该有一名合格人员在需要时操作飞机制动。应额外指派人员监视翼尖和机尾是否与其他物体有足够的间隙。

大多数飞机的前起落架都有一个非常明确的限制，它可以被控制。当牵引时，很容易超过这个限制。如果超过转弯半径，将损坏前起落架支柱和转向机构。损坏程度可能相当大，需要更换前起落架支柱。一些飞机具有断开锁定装置的方法，因此前轮可以旋转以方便机动。如果是这种情况，当飞机被拖曳时，锁定装置必须始终断开。此外，记得从飞机上卸下拖杆后重新锁定。当拖杆连接在飞机上时，乘坐飞机的人不应试图操纵前轮。

异物损坏（FOD）

如果螺母、螺栓、安全线等异物被拉进涡轮发动机的进气道，或者通过旋转螺旋桨叶片的电弧线，则很容易造成损坏，从而导致灾难性的故障。此外，如果这些物体被困在高速排气发动机中并撞击其他物体，这些物体可能会遭受严重损害。因此，保持机场航线的清洁是极其重要的。此外，作为一名飞机维修技术员，应该养成一个习惯，即把在斜坡上找到的所有松动的硬件和抹布捡起来，并把它们存放在一个合适的容器里。

人身安全比设备损坏重要得多。因此，要注意螺旋桨和喷气进气道使飞机操作区域变得极其危险。良好的维护实践要求飞机运行发动机时必须点亮导航灯和旋转信标。此外，应派第二个人驻扎在地面上，警示该地区的其他人。如果有必要接近一架发动机还在运转的飞机，一定要引起驾驶舱人员的注意，这样他们就能随时知道你在哪里。螺旋桨的危险是如此明显，几乎不需要说，但可悲的是，许多人因为误入旋转的螺旋桨而丧生。始终要当心旋转的螺旋桨。

涡轮喷气飞机的前部和后部都很危险。发动机在运转过程中，会扰动大量的空气，在进气道前方形成低压区（图3-23）。这个低压区域可以将人吸入发动机，当然也可以吸入帽子、写字板和宽松的衣服等物品。在飞机后方的排气区，高速的废气排放会对人员和设备造成严重的伤害。

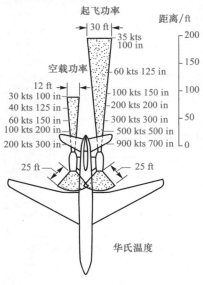

图 3-23　远离危险区域

压缩气体安全

所有飞机维修车间都有压缩气体。例如，压缩空气驱动风钻发动机、铆钉枪、喷漆枪和清洁枪。此外，压缩氮气用于给轮胎和减振装置充气，而压缩乙炔用于焊接。

大多数车间压缩空气保存在储罐中，并通过高压管线输送到整个车间。这种高压会造成严重的伤害。例如，如果一股浓缩的压缩空气流吹过皮肤上的伤口，空气就有可能进入血液中，对人造成严重的伤害或导致死亡。使用压缩空气时要非常小心，不要把灰尘或碎屑吹到附近任何人的脸上。

使用气动工具时，应佩戴护目镜，防止眼睛受伤。为了防止软管破裂造成伤害，请始终保持空气软管和配件处于良好状态。

太多的事故发生在给轮胎充气或放气的时候。始终使用经过校准的轮胎量规，并确保使用处于良好工作状态的调节器。

高压压缩气体如果处理不当，尤其危险。氧气和氮气通常储存在航空维修车间的钢瓶中，压力约为 2 000 psi（1 psi=6.895 kPa），这些钢瓶上装有黄铜阀门。如果一个钢瓶被撞翻，阀门断开，逸出的高压气体就会像火箭一样推动容器。由于这会造成很大的危险，因此应该确保所有的气瓶都有适当的支撑，在储存中固定高压钢瓶的一种常用方法是将它们拴在建筑物上。此外，阀盖应该安全地安装在没有连接系统的任何钢瓶上，这可以保护阀门免受损坏。

氧气瓶要特别小心地处理，这是极其重要的。氧气除具有其他高压气体所固有的所有危险外，还有爆炸的危险。例如，你绝不能让氧气接触石油产品，如油或油脂，因为氧气会导致这些材料自燃和燃烧。此外，千万不要使用油性抹布以及油性或油腻的工具在氧气瓶上安装配件或调节器。

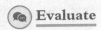 **Evaluate**

任务名称		地勤服务基本知识	姓名	组别	班级	学号	日期	
考核内容及评分标准				分值	自评	组评	师评	平均分
三维目标	知识	了解地勤服务的相关知识		25 分				
	技能	能阅读专业英语文章、翻译专用英语词汇		40 分				
	素养	在学习过程中秉承科学精神、合作精神		35 分				
加分项	收获（10 分）	收获（借鉴、教训、改进等）：	你进步了吗？			加分		
			你帮助他人进步了吗？					
	问题（10 分）	发现问题、分析问题、解决方法、创新之处等：				加分		
总结与反思						总分		

 Unit 3　Oil Industry

Unit objectives

【Knowledge goals】
1．掌握定量测井解释的相关知识，其中包括储层参数、地层电阻率、储集岩传导率等。
2．掌握定向井、斜井钻井装置及其钻井方法等。

【Skill goals】
1．能够阅读专业英语文章。
2．可以翻译专用英语词汇。

【Quality goals】
1．在学习过程中发扬科学研究精神。
2．增强团队合作意识。

3.1　Fundamentals of Quantitative Log Interpretation

 Text

石油行业

I . Reservoir Parameters to be Evaluated

Almost all oil and gas produced today comes from accumulations in the pore spaces of reservoir rocks. Although oil shales and tar sands may someday become important petroleum sources, they will be considered only briefly in this module.

The amount of oil or gas contained in a unit volume of the reservoir is the product of its porosity by the hydrocarbon saturation. Porosity is the pore volume per unit volume of formation. Hydrocarbon saturation is the fraction (or percentage) of the pore volume filled with hydrocarbons.

In addition to the porosity and the hydrocarbon saturation, the volume of the formation containing hydrocarbons is needed in order to determine if the accumulation can be considered commercial. Knowledge of the thickness and area of the reservoir are needed for computation of its volume.

To evaluate the producibility of a reservoir, it is useful to know how easily fluid can flow through the pore system. This property of the formation, which depends on the manner in which the pores are interconnected, is its permeability. The main physical parameters needed to evaluate a reservoir, then, are its porosity, hydrocarbon saturation permeable bed thickness, and permeability. These parameters can be derived or inferred from electrical, nuclear, and acoustic logs.

This module is concerned mainly with the determination of porosity and water saturation. It also explains how logs are used to obtain valuable information about permeability, lithology, and producibility, and to distinguish between oil and gas.

Of the formation parameters obtained directly from logs, resistivity is of particular importance. It is essential to saturation determinations. Resistivity measurements are used, singly and in combination, to deduce formation resistivity in the uninvaded formation, i.e., beyond the zone contaminated by borehole fluids. It is also used to determine the resistivity close to the borehole, where mud filtrate has largely replaced the original fluids. Resistivity measurements, along with porosity and water resistivity, are used to obtain values of water saturation. Saturation values from both shallow and deep resistivity measurements are compared in order to evaluate the producibility of a formation.

Several different logs may be used to determine porosity: sonic, formation density, and neutron logs have responses that depend primarily on formation porosity. They are also affected by rock properties, each in a different way, so combinations of two or three of these logs yield better knowledge of the porosity, lithology, and pore geometry, also, they will frequently distinguish between oil and gas.

Permeability, at the present time, can only be estimated from empirical relationship. These estimates should be considered as having only order-of-magnitude accuracy.

II. Resistivity

The resistivity of a substance is its ability to impede the flow of electric current through that substance. The resistivity unit used in electrical logging is the $\Omega \cdot m^2/m$, usually written $\Omega \cdot m$. The resistivity of a formation (in $\Omega \cdot m$) is the resistance of a one-meter cube when the current flows between opposite faces of the cube.

Electrical conductivity is the reciprocal of resistivity, expressed in mhos/m. In electrical logging practice, to avoid decimal fractions, conductivity is expressed in millimhos per meter (mmho/m). A resistivity of 1 $\Omega \cdot m$ corresponds to a conductivity of 1,000 mmho/m; 100 $\Omega \cdot m$ corresponds to 10 mmho/m, etc.

Formation resistivities usually fall in the range from 0.2–1,000 $\Omega \cdot m$. Resistivities higher than 1,000 $\Omega \cdot m$ are uncommon in permeable formations.

III. Metallic Conduction

Logs are sometimes used to locate and evaluate ore bodies. Many ores, such as galena, chalcopyrite, etc., have very high conductivities. Their depth and thickness may be readily determined from resistivity logs run in test borings.

IV. Conduction in Reservoir Rocks

Most formations logged for oil and gas are made up of rocks which when dry will not conduct electrical current. Current flows through the interstitial water, made conductive by salts in solution. These salts dissociate into positively charged cations (Na^+, Ca^{2+}...) and negatively charged anions (Cl^-, SO^{2-}...). Under the influence of an electrical field these ions move, carrying an electrical current through the solution. Other things being equal, the greater the salt concentration, the lower the resistivity of the formation water, hence of the formation.

In the case of formation water containing sodium chloride only, its resistivity, R_w, is a function of its salinity and temperature. Knowing these two parameters, one can compute R_w from related data charts. For a water containing other salts, in addition to sodium chloride, R_w can be estimated from its chemical analysis.

V. Effect of Temperature

In quantitative log interpretation it is necessary to correct all fluid resistivities to formation temperature. Knowing bottom-hole temperature, one can estimate formation temperature from related data charts. These charts are based on the assumption that temperature increases linearly with depth. This condition may not always be met in practice.

VI. Shale Conduction

Shaliness also contributes to formation conductivity. Shale conduction differs from electrolytic conduction described above in that the current is not carried by ions moving freely in a solution. Rather, conduction is an ion-exchange process whereby (usually the positively charged) ions move under the influence of the impressed electric field between exchange sites on the surfaces of the clay particles.

Surface conductance at the shale-liquid interfaces is an important factor in the effect of shaliness on conductivity, and its influence is often disproportionate to the quantity of shale. The net effect of shaliness depends on the amount, type, and distribution of the shale, and on the nature and relative amount of the formation water.

VII. Formation Factor and Porosity

It has been established experimentally that the resistivity of a clean formation (i.e., one containing no appreciable amount of clay) is proportional to the resistivity of the brine with which it is fully saturated. The constant of proportionality is called formation resistivity factor, (F). Thus, if R_0 is the resistivity of a nonshaly formation sample 100% saturated with brine of resistivity R_w, then

$$F = \frac{R_0}{R_w} \qquad (3-1)$$

(Throughout this module, the term "formation water" will be used to identify the water which existed in the formation before it was penetrated by the drill bit. In most cases it is the same as "connate water".)

For a given porosity the ratio R_0/R_w remains nearly constant for all values of R_w below about 1 Ω · m. However, experiments show that in more resistive waters, the value of F is reduced as R_w rises, and as the grain size of the sand decreases. This phenomenon is attributed to a greater proportionate influence of the surface conductance of the grains, in fresher water.

The porosity (ϕ) of a rock is the fraction of the total volume occupied by pores or voids. Formation factor is a function of porosity, and also of pore structure and pore-size distribution.

Archie formula:

$$F = a/\phi^m \quad (3-2)$$

where m is the cementation factor; the constant a is determined empirically.

Satisfactory results are usually obtained with, in sands:

$$F = \frac{81}{\phi^2} \quad (3-3a)$$

In compacted formations:

$$F = \frac{1}{\phi^2} \quad (3-3b)$$

Within their normal ranges of application these two formulas differ little from the so-called "Humble formula":

$$F = \frac{0.62}{\phi^{2.15}} \quad (3-3c)$$

This is the relationship used in the Schlumberger charts, unless otherwise specified.

While the Humble formula is satisfactory for sucrosic rocks better results are obtained using $F = \frac{1}{\phi^2}$ in chaiky rocks and $F = \frac{1}{\phi^{2.2-2.5}}$ in compact or oolicastic rocks. In some oolicastic rocks, m may be as high as 3.0.

VIII. Water Saturation

In a formation containing oil or gas, both of which are electrical insulators, the resistivity is a function not only of F and R_w, but also of the water saturation, S_w. S_w is the fraction of the pore volume occupied by formation water ($1-S_w$) is the fraction of the pore volume occupied by hydrocarbons. Archie determined experimentally that the water saturation of a clean formation can be expressed in terms of its true resistivity, R_t, as:

$$S_w^n = \frac{FR_w}{R_t} \quad (3-4)$$

where n is the saturation exponent, generally equal to 2; FR_w is equal to R_0, the resistivity of the formation when 100% saturated with water of resistivity R_w. The equation may then be written as:

$$S_w = \sqrt{\frac{R_0}{R_t}} \qquad (3-5)$$

The earliest quantitative interpretations used this formula based on resistivity only. Its use assumed that the permeable formation had the same formation factor in the water-bearing interval of the bed (where R_0 was determined) as in the hydrocarbon-bearing interval (where R_t was determined). The ratio R_t/R_0 was called the "resistivity index".

The above formulas are good approximations in clean formations having a fairly regular distribution of the porosity (intergranular or inter-crystalline porosity). In formations with fractures or vugs, the formulas can still be used, but the accuracy is not good.

IX. Invasion

During the drilling operation, the mud in the borehole is usually conditioned so that the hydrostatic pressure of the mud column is greater than the pressure of the formations. The differential pressure forces mud filtrate into the permeable formations, and the solid particles of the mud are deposited on the borehole wall where they form a mud cake. Mud cake usually has very low permeability and considerably reduces the rate of infiltration.

Very close to the hole, all the formation water and some of the hydrocarbons, if present, are flushed away by the filtrate. The resistivity, R_{x0}, of this "flushed zone" is expressed by the Archie formula as:

$$R_{x0} = \frac{FR_{mf}}{S_{x0}^2} \qquad (3-6)$$

where R_{mf} is the resistivity of the mud filtrate; S_{x0} is the mud filtrate saturation. S_{x0} is equal to $(1-S_{hr})$, S_{hr} being the residual hydrocarbon saturation in the flushed zone. S_{hr} depends to some extent on the hydrocarbon viscosity, generally increasing as the viscosity increases.

Fig.3-24 to Fig.3-26 show the horizontal profile of the permeable water-bearing formation near the hole, the radial distribution of resistivity around the hole, and the radial distribution of fluid near the hole during the drilling operation, respectively.

Farther out from the borehole, the displacement of formation fluids is less and less complete, resulting in a transition zone with a progressive change in resistivity from R_{x0} to the resistivity R_t of the uninvaded formation (Fig. 3-27). Sometimes, in oil- or gas-bearing formations, where the mobility of the hydrocarbons is greater than that of the water due to relative permeability differences, the oil or gas moves away faster than the interstitial water. In this case, there may be formed between the flushed zone and uninvaded zone an annular zone with a high formation water saturation; if R_{mf} is greater than R_w, this annulus will have a resistivity lower than either R_{x0} or R_t. Annuli do not occur in all oil-bearing formations, and when they do, they generally disappear with time.

In fractured formations the mud filtrate goes easily into the fissures, but penetrates very little into the unfractured blocks of low-permeability matrix. Therefore only a small proportion of the original fluid is displaced by the filtrate even very close to the borehole. R_{x0} then, does not differ much from R_t, and the Archie relationship as expressed in Eq. 3-4 is not applicable.

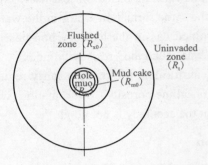

Fig. 3-24 Horizontal section through a permeable water-bearing BED

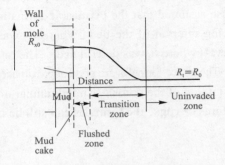

Fig. 3-25 Redial distribution of resistivity ($R_{mf} \gg R_w$, water-bearing BED)

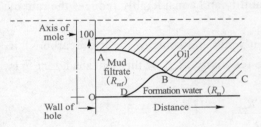

Fig. 3-26 Redial distribution of fluids in the vicinity of the borehole, oil-bearing BED (qualitative)

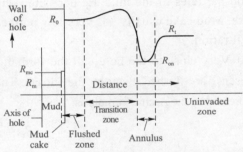

Fig. 3-27 Radial distribution of resistivities ($R_{mf} \gg R_w$, oil-bearing BED, $S_w \ll 50\%$)

【Words and phrases】

I.

accumulation	n. 积累；堆积；堆积物，堆积量
reservoir	n. 水库；蓄水池；储藏；蓄积；（机器等的）储液器
reservoir rock	储集岩
shale	n. [岩] 页岩；泥板岩
porosity	n. 有孔性；多孔性
hydrocarbon	n. [有化] 碳氢化合物
hydrocarbon saturation	含油气饱和度
fraction	n. 分数；小数；小部分；微量；持不同意见的小集团
pore	n. （皮肤上的）毛孔；（植物的）气孔

	v. 仔细阅读；认真钻研
interconnect	vt. 使互相连接
	vi. 互相联系
permeable	adj. 能透过的；有渗透性的
permeability	n. 渗透性；透磁率，磁导率；弥漫
lithology	n. 岩石学；岩性学；岩性，岩石特征
resistivity	n.［电］电阻率；抵抗力；电阻系数
be concerned with	关注；从事于；涉及
mud	n. 泥；淤泥，泥浆；（多指与腐化有关的）侮蔑；诽谤
filtrate	n. 滤液；滤气
	v. 过滤；筛选
mud filtrate	泥浆滤液
sonic	adj. 声速的；声音的；音波的
formation density	地层密度
pore geometry	孔隙几何形状
empirical	adj. 经验主义的；以经验为依据的
empirical relationship	［数］经验关系

Ⅱ.

millimho	n. 毫欧姆（电导、导纳和电纳单位）

Ⅲ.

interval	n. 间隔；间隙；幕间休息；中场休息；音程
resistivity index	电阻率系数；电阻增大率
approximation	n. 粗略估计；近似值；类似事物；相似物；近似法
intergranular	adj. 颗粒间的；晶粒间的
fracture	n. 破裂；断裂；骨折；（尤指岩层的）裂缝；裂痕
vug	n.［晶体］晶簇

Ⅳ.

hydrostatic	adj. 流体静力学的；静水力学的
hydrostatic pressure	静水压力
differential pressure	压差
be equal to	等于
flushed	adj. 脸红的；自豪或激动而兴奋的
	v. 使激动

flushed zone　　　　　　　　　　　冲洗区
invasion　　　　　　　　　　　　　n. 侵略；入侵；涌入；大批进入；侵犯；
　　　　　　　　　　　　　　　　　干预；侵袭；扩散
profile　　　　　　　　　　　　　　n. 侧面；剖面
unfractured　　　　　　　　　　　 adj. 未断裂的；未挫伤的

Course practice

1. Complex sentence analysis.

（1）Porosity is the pore volume per unit volume of formation.

per：每，每一；按照，根据。

【译文】孔隙度是单位体积地层的孔隙体积。

（2）This module is concerned mainly with the determination of porosity and water saturation.

is concerned with：和……相呼应；与……合作；和……一致。

【译文】本模块主要关注孔隙度和含水饱和度的测定。

（3）It is essential to saturation determinations.

It is essential to：至关重要的是……；……是必不可少的。

【译文】它对饱和度测定是必不可少的。

（4）Formation resistivities usually fall in the range from 0.2 to 1 000 Ω·m.

range from …to …：范围从……到……

【译文】地层电阻率通常为 0.2~1 000 Ω·m。

（5）Most formations logged for oil and gas are made up of rocks which when dry will not conduct electrical current.

be made up of：由……组成。

【译文】大多数储存石油和天然气的地层都是由岩石组成的，这些岩石在干燥时不会传导电流。

（6）In the case of formation water containing sodium chloride only, its resistivity, R_w, is a function of its salinity and of temperature.

In the case of：表示在特定情况或条件下（在……情况下）。

【译文】对于仅含氯化钠的地层水，其电阻率 R_w 是其盐度和温度的函数。

（7）In quantitative log interpretation it is necessary to correct all fluid resistivities to formation temperature.

it is necessary to do：做……是必要的。

【译文】在定量测井解释中，必须校正所有流体对地层温度的电阻率。

2. Translate the following paragraphs.

（1）In water-bearing zones, only two liquids are present, filtrate and formation water. With relatively fresh muds, the filtrate is less dense, and will move upward toward the boundary of the permeable bed.

(2) Extreme cases have been observed in which the invaded zone has disappeared in the lower part of the bed.

3. Choose the proper answer to fill in the blank and translate the sentences.

(as, in, to, on, from, by, at, of)

(1) When oil is present (   ) the movable liquid, two general patterns are possible.

(2) The quantity of filtrate (   ) the system is effectively fixed.

(3) Porosity values can be obtained (   ) a sonic log, a formation density log.

(4) In addition (   ) porosity, these logs are affected (   ) other parameters, such as lithology, nature of the pore fluids, and shaliness.

Translation of the text

定量测井解释基础

I. 评价的储层参数

目前几乎所有的石油和天然气都来自储层岩石孔隙中的堆积物。虽然油页岩和沥青砂可能在某一天会成为重要的石油来源，但本模块只对它们作简要的讨论。

单位体积储层中所含的石油或天然气的量是其孔隙度与碳氢化合物饱和度的乘积。孔隙度是单位体积地层的孔隙体积。碳氢化合物饱和度是指孔隙体积中充满碳氢化合物的分数（或百分比）。

除孔隙度和碳氢化合物饱和度外，还需要确定含碳氢化合物地层的体积，以确定该储层是否具有商业价值。要计算储层的体积，需要知道储层的厚度和面积。

为了评价储层的产能，需要了解流体通过孔隙系统的容易程度。储层的这种性质取决于孔隙相互连接的方式，即渗透率。因此，评价储层所需要的主要物理参数是孔隙度、碳氢化合物饱和度、渗透层厚度和渗透率。这些参数可以从电测井、核测井和声波测井中获得或推断出来。

本模块主要关注孔隙度和含水饱和度的测定，还解释了如何使用测井来获取有关渗透率、岩石特性和产能等有价值的信息，以及如何区分石油和天然气。

在直接从测井资料中获得的地层参数中，电阻率尤为重要。它对饱和度的测定是必不可少的。电阻率测量可以单独使用，也可以组合使用，以推断未侵入地层的地层电阻率，即超出井内流体污染的区域。它还用于确定井眼附近的电阻率，在那里泥浆滤液已经大量取代了原始流体。通过对电阻率，以及孔隙度和水电阻率的测量，可获得含水饱和度的值。通过比较浅层和深层电阻率测量的饱和度值来评估地层的产能。

几种不同的测井可用于确定孔隙度，声波测井、地层密度测井和中子测井的响应主要取决于地层孔隙度。它们还受到岩石性质的影响，每种性质都有不同的影响方式，因此，将 2～3 种测井资料组合在一起，可以更好地了解孔隙度、岩性和孔隙几何形状，此外，还经常用它们来区分石油和天然气。

目前，渗透率只能通过经验关系来估计，这些估计应该被认为只有数量级的准确性。

Ⅱ. 电阻率

一种物质的电阻率是它阻碍电流通过该物质的能力。电测井中使用的电阻率单位为 $\Omega \cdot m^2/m$，通常写成 $\Omega \cdot m$。储层的电阻率是指当电流在立方体的相对面之间流过时，1 m^3 的电阻（以 $\Omega \cdot m$ 为单位）。

电导率是电阻率的倒数，单位是 mho/m。在电测井实践中，为了避免出现小数，电导率以毫欧姆/米（mmho/m）表示。1 $\Omega \cdot m$ 的电阻率对应 1 000 mmho/m 的电导率；100 $\Omega \cdot m$ 的电阻率对应 10 mmho/m 等。

地层电阻率通常为 0.2～1 000 $\Omega \cdot m$。在渗透性地层中，电阻率高于 1 000 $\Omega \cdot m$ 的情况并不多见。

Ⅲ. 金属传导

测井有时被用来定位和评价矿体。许多矿石，如方铅矿、黄铜矿等，都具有很高的电导率，它们的深度和厚度可以很容易地从测试钻孔的电阻率测井中确定。

Ⅳ. 储集岩的传导

大多数储存石油和天然气的地层都是由岩石组成的，这些岩石在干燥时不会传导电流。电流流经间隙水，溶液中的盐使其具有导电性，这些盐解离成带正电的阳离子（Na^+、Ca^{2+} 等）和带负电的阴离子（Cl^-、SO_4^{2-} 等）。在电场的作用下，这些离子移动，携带电流通过溶液。在其他条件相同的情况下，盐浓度越高，地层水的电阻率就越低，因此地层的电阻率也就越低。

对于仅含氯化钠的地层水，其电阻率 R_w 是其盐度和温度的函数。知道了这两个参数，就可以从相关图表中计算 R_w。对于含有除氯化钠以外的其他盐类的水，R_w 可以通过化学分析来估计。

Ⅴ. 温度的影响

在定量测井解释中，必须校正所有流体对地层温度的电阻率。知道了井底温度，就可以根据相关图表估算地层温度。这些图表是基于温度随深度线性增加的假设，这个条件在实践中可能并不总能得到满足。

Ⅵ. 页岩传导

泥质也有助于地层的导电。页岩传导不同于上述的电解传导，因为电流不是由在溶液中自由移动的离子携带的。更确切地说，传导是一种离子交换过程，即离子（通常是带正电荷的离子）在外加电场的影响下在黏土颗粒表面的交换点之间移动。

页岩-液体界面的表面电导率是对页岩质电导率产生影响的重要因素，其影响往往与页岩的数量不成比例。页岩的净效应取决于页岩的数量、类型和分布，以及自然地层水的性质和相对量。

VII. 地层因素和孔隙度

试验证明，干净地层（即不含明显黏土的地层）的电阻率与完全饱和的盐水的电阻率成正比。比例常数称为地层电阻率系数（F）。因此，若 R_0 为非泥质地层样品电阻率，100% 饱和盐水的电阻率为 R_w，则

$$F = \frac{R_0}{R_w} \qquad (3-1)$$

（在本模块中，术语"地层水"将用于识别在钻头穿透地层之前存在于地层中的水。在大多数情况下，它与"天然水"相同。）

对于给定的孔隙度，若 R_w 低于约 1 Ω·m 的所有值，R_0/R_w 的比值几乎保持不变。然而，试验表明，在阻力较大的水域中，F 值随着 R_w 的增大和砂粒尺寸的减小而减小。这一现象归因于淡水中颗粒表面电导率的比例影响更大。

岩石的孔隙度（ϕ）是孔隙或空隙占总体积的比例。地层因子是孔隙度的函数，也是孔隙结构和孔径分布的函数。

阿尔奇公式：

$$F = a/\phi^m \qquad (3-2)$$

式中，m 为胶结因子；常数 a 是根据经验确定的。

通常在砂层中：

$$F = \frac{81}{\phi^2} \qquad (3-3a)$$

在压实地层中：

$$F = \frac{1}{\phi^2} \qquad (3-3b)$$

在它们的正常应用范围内，这两个公式与"汉布尔公式"差别不大：

$$F = \frac{0.62}{\phi^{2.15}} \qquad (3-3c)$$

除非另有说明，这是斯伦贝谢图表中使用的关系。

汉布尔公式对于蔗糖质岩石具有人满意的结果，在白垩质岩石中使用 $F=\frac{1}{\phi^2}$；在致密或鲕粒岩石中使用 $F=\frac{1}{\phi^{2.2-2.5}}$，在一些特殊的鲕粒岩石中，$m$ 可高达 3.0。

VIII. 含水饱和度

含有石油或天然气的地层都是电绝缘体，其电阻率不仅与 F 和 R_w 有关，还与

含水饱和度 S_w 有关。S_w 为地层水占孔隙体积的比例，（$1-S_w$）为石油和天然气占孔隙体积的比例。阿尔奇通过试验确定干净地层的含水饱和度可以用其真实电阻率 R_t 表示为

$$S_w^n = \frac{FR_w}{R_t} \tag{3-4}$$

式中，通常取饱和指数 n 为 2；FR_w 等于 R_0，即地层 100% 饱和时电阻率为 R_w 的水的电阻率。于是，式（3-4）可以写成

$$S_w = \sqrt{\frac{R_0}{R_t}} \tag{3-5}$$

最早的定量解释仅使用基于电阻率的公式。使用该方法的前提是渗透率在地层含水层段（确定 R_0 值）与含油层段（确定 R_t 值）具有相同的地层因子。比值 R_t/R_0 称为"电阻率指数"。

式（3-5）在孔隙度（晶间或晶体孔隙度）分布规则的干净地层中相当近似。在有裂缝或孔洞的地层中，仍然可以使用这些公式，但精度不高。

IX. 扩散

在钻井作业中，通常对井内泥浆进行调节，使泥浆柱的静水压力大于地层压力。压差迫使泥浆过滤到渗透性地层中，泥浆的固体颗粒沉积在井壁上，形成泥饼。泥饼通常渗透性很低，大大降低了渗透速度。

在非常靠近井眼的地方，所有地层水和一些碳氢化合物（如果存在）都被滤液冲走。该"冲刷带"的电阻率 R_{x0} 用阿尔奇公式表示为

$$R_{x0} = \frac{FR_{mf}}{S_{x0}^2} \tag{3-6}$$

式中，R_{mf} 为泥浆滤液电阻率；S_{x0} 为泥浆滤液饱和度。$S_{x0} = (1-S_{hr})$，S_{hr} 为冲蚀带残余烃饱和度。S_{hr} 在一定程度上取决于烃的黏度，通常随着黏度的增加而增加。

图 3-24 ～ 图 3-26 分别展示了钻井作业过程中，在靠近井眼区域内，可渗透含水地层的水平剖面、井眼周边区域电阻率径向分布、井眼附近流体径向分布状况。

离井越远，地层流体的位移越小，形成了一个过渡带，电阻率从 R_{x0} 到未侵地层的电阻率 R_t 逐渐变化（图 3-27）。有时，在含油或含气地层中，由于相对渗透率的差异，碳氢化合物的流动性大于水的流动性，石油或天然气的流动速度比间隙水高。这种情况下，在冲刷带和未侵入带之间可能形成一个地层含水饱和度较高的环空带；如果 R_{mf} 大于 R_w，则该环空带的电阻率将低于 R_{x0} 或 R_t。环空带并不会出现在所有含油地层中，当它们出现时，通常会随着时间的推移而消失。

在裂缝性地层中，泥浆滤液很容易进入裂缝，但很少渗透到低渗透基质的未裂缝块体中。因此，即使非常接近井眼，也只有一小部分原始流体被滤液置换。因此，R_{x0} 与 R_t 相差不大，不适合式（3-6）所示的阿尔奇关系。

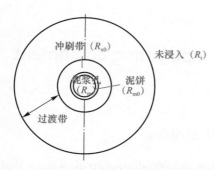

图 3-24 通过可渗透含水地层的水平剖面

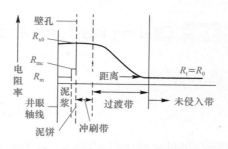

图 3-25 电阻率（$R_{mf} \gg R_w$、含水 BED）径向分布

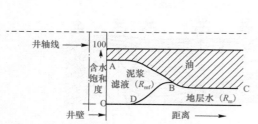

图 3-26 井眼附近流体径向分布，含油 BED（定性）

图 3-27 电阻率径向分布（$R_{mf} \gg R_w$，含油 BED，$S_w \ll 50\%$）

Evaluate

任务名称		定量测井解释基础	姓名	组别	班级	学号	日期	
考核内容及评分标准				分值	自评	组评	师评	平均分
三维目标	知识	了解定量测井解释的相关知识，其中包括储层参数、地层电阻率、储集岩传导率等		25 分				
	技能	能阅读专业英语文章、翻译专用英语词汇		40 分				
	素养	在学习过程中秉承科学精神、合作精神		35 分				
加分项	收获（10 分）	收获（借鉴、教训、改进等）：		你进步了吗？			加分	
				你帮助他人进步了吗？				
	问题（10 分）	发现问题、分析问题、解决方法、创新之处等：					加分	
总结与反思							总分	

3.2 Drilling Platform and Drill Pipe Lifting Equipment

Text

I. Directional Well Drilling

A directional well drilling refers to a well that specifically points to a point that is "vertically projected away from its mouth". Directional drilling is currently widely used in oil, natural gas, and solid mineral drilling (Fig. 3-28).

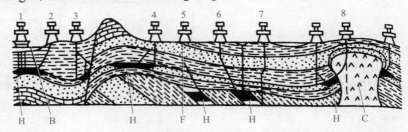

Fig. 3-28 Example of directional drilling
1—Submarine excavation; 2—Offshore oil fields erupt from the shore;
3—The wellbore deviates from the discharge zone (fracture zone) towards the oil zone;
4—When the working face is located below the unable area to install drilling rig, drill a directional shaft;
5—Drilling monoclinic oil layers; 6—Drilling auxiliary inclined shafts to extinguish fires or cover fountains;
7— Lateral movement in case of accidents; 8— Drilling directional shafts in salt dome enclosed areas;
H—Petroleum; B—Water; F—Gas; C—salt

There are two drilling methods as below.

Rotor drilling The rotor drilling represents the intermittent process of continuous cutting (lateral) to bend the wellbore. Bottom hole engine ensures the continuous bending process of the wellbore.

Rotary drilling In Russia, the vast majority of directional drilling is done using bottom hole engines, while in other countries, rotary drilling is dominant, and bottom hole engines are mainly used in curvature groups in the specified direction. Domestic and foreign experts believe that the curvature set of screw bottom hole engines in a given direction is the most promising. These engines have higher power than turbo drills and lower shaft speeds, which are beneficial for curvature setting.

The Profile of the Directional Oil Well

The profile of the directional oil well should minimize the funding and time required for drilling to ensure completion of the drilling task.

In directional drilling, the four types of profiles (Fig. 3-29) are the most widely distributed.

Section I [Fig. 3-29(a)] is the most common. It consists of three parts: the first part is the vertical area, the second part is the float curve, and the third part is the float line. This type of profile is mainly used for drilling single-layer deposits with large deviations at the average SCN deposit depth.

Section II [Fig. 3-29 (b)] consists of four parts: the upper part of the vertical second part is made up of the rising curvature curve of the third part-directional straight line, and the fourth part is made up of the falling curvature curve. Usually, this configuration file is used in a slightly modified form — without region 3, that is, region 2 with an increase in curvature followed by region 4 with a decrease in curvature. This type of profile is usually used for drilling directional wells up to a depth of 2,500 m.

Section III [Fig. 3-29 (c)] is smaller than the first two. It consists of two parts: the first part is vertical, and the second part is drawn along a curve that gradually increases the inclination angle of the barrel. When it is necessary to maintain the predetermined angle of the wellbore entering the formation, drilling can be carried out according to this profile.

Section IV [Fig. 3-29 (d)] is used for deep directional drilling. This profile is different from previous profiles in that a curve section 4 is added to vertical profile 1, profile 2 drawn along the curve, and profile 3 representing an directional straight line. Its feature is a decrease in curvature obtained, which means that the barrel is peeled off to the vertical and straight profile 5 is added to the vertical profile. When there are multiple production layers at the bottom of the oil well, this type of profile should be used.

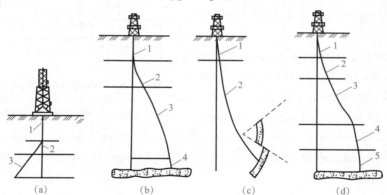

Fig. 3-29 Directional shaft profile
(a) Section I ; (b) Section II ; (c) Section III ; (d) Section IV
1—Vertical area; 2—The part with increased curvature; 3—Directional straight line;
4—Curve descent section; 5—Vertical section

The above profile is a curve located on the same vertical plane. These configuration files are called regular configuration files. In drilling, it is sometimes necessary to use a profile, which is a spatial curve similar to a helix or spatial helix, i.e., a spatial profile. This type of oil well is drilled in areas where geological conditions have a significant impact on the spontaneous bending of the wellbore. When constructing the profiles of these oil wells, the spontaneous bending mode of the wells should be utilized as much as possible to minimize the deviation drilling interval. In Russia, directional drilling of spatial profile is common in Grozny oil region.

Ⅱ. Inclined Well Drilling

After the preparation work is completed, start assembling the deflection layout. The layout is assembled according to the drilling work plan, including the drill bit, bottom hole engine, deflection device (usually curve converter), and then lowered into the well for drilling. In order to know the position of the deflector, the marked position on each connecting tube will be fixed when placing the instrument. The distance between labels is determined by a metal tape measure or other most common method-using paper tape, which is a thick strip of paper with a width of 8 – 10 cm and a length equal to or slightly greater than the length of the drill pipe lock circle. The stripes of the paper are folded in half and marked with o (deflector) in the middle of its length.

Add an o mark on the curve converter, an o mark on the joint, and mark y on the paper tape (Fig. 3-30). After connection, the y mark on the paper tape is merged with the y mark on the УБТ coupling. Draw mark 1 on the marking on the drill pipe joint on the paper tape and place the drilling tool connector in the well with the length of the drill pipe.

Tighten the second pipe and secure it, combine the mark 1 on the belt with the mark on the clutch of the first pipe, mark 2 on the joint of the second pipe on the belt,

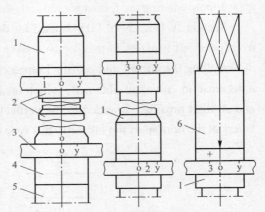

Fig. 3-30　Directional drill pipe lowering
1—Drill pipe; 2—УБТ; 3—Paper tape;
4—Translator; 5—Engine base; 6—Main drill pipe

and cross off the previous pipe. Therefore, there is no paper tape recording the distance between all descending drill pipe labels. After all drill pipes are lowered, tighten the main drill pipe.

Mark 3 on the last drill pipe belt with the mark on the pipe joint, and use the main drill pipe (square drill pipe) to mark the o mark indicating the direction of movement of the inclinometer on the adapter.

In geological logging data, the input bridge direction φ_M and the bottom hole deflection design azimuth angle $\varphi_{пр}$. In order to install the deflector in the desired direction, a difference $\lambda=\varphi_{пр}-\varphi_M$ is defined. The resulting angle λ is placed on the circle of the rotor table in a clockwise or counterclockwise direction from the bridge, and a mark П is placed to indicate the direction towards the design point, depending on the symbol. Rotate the drill pipe with the rotor and combine the o mark on the main drill pipe rotor with the o mark on the fixed part of the rotor table. At this position, use a template to move one of the square edges onto the rotor with chalk. Usually, the most suitable edge for observation. Place the tools in the appropriate position, remove the elevator, and then rinse them down to the working surface. After reaching the working surface, the deflector should point in the specified direction, and the o mark on the main drill pipe rotor should be the same as the o mark on the rotor (Fig. 3-31).

During the lifting process, the drill pipe is installed on the fingers in the same order as the drilling process. During the drilling process, marks are filled on the newly lowered pipeline, and the deflector positions ω of the drill pipe in the manner described above, depending on the reaction torque of the drilling engine and the length of the drill pipe. Generally, in actual work, the thread angle ω of 168 mm and 140 mm drill pipes per 100 m is 3° and 5° (the length of drill pipe does not exceed 1,000 -1,500 m).

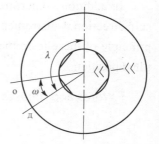

Fig. 3-31 Rotor deflection direction diagram

Deflection Direction

The drill pipe can be lowered into an inclined well with a working surface slope greater than 3°, just like drilling a regular vertical well. In this case, an electromagnetic compass and a inclinometer with a magnetic converter are used to point the working surface deflector in the correct direction. The drilling positioning of the magnetic compass and magnetic converter inclinometer deflector is based on a multi-point inclinometer. Use an electromagnetic compass to measure orientation. This tool includes a drill pipe string that ends in a diamagnetic tube (1×18N9T steel or D16T aluminum alloy) at the bottom. Tighten the converter at the bottom of the diamagnetic tube to secure the magnetic field source in it. Magnetic converter is connected to limit converter and curve converter, with a cross connection between the latter. Next is the engine.

After the tool is lowered to the working surface, tighten the converter and the rotating sleeve, and fix the logging roller in the upper pipe clutch. Lower the inclinometer with an electromagnetic compass inside the drill pipe on cable. Premeasure the azimuth angle of the wellbore in the diamagnetic tube above the magnetic converter. The position of the eccentric is determined through point sampling in the magnetic converter.

Under the guidance of measurement data, the rotor rotates the pipeline to the desired deflector position, and then recalibrates the measurement point to check if the deflector is correctly installed. Then take out the inclinometer from the drill pipe, mark the position of the instrument, tighten the main drill pipe (square), fix one of the ribs, lock the rotor, and start drilling.

Deflection surface orientation:

$$\beta = 360° - \Delta + \varphi$$

where Δ is the inclinometer reading (°) measured by the magnetic converter; φ is the azimuth angle of the wellbore when measured in a diamagnetic tube (°).

The bottom hole positioning method of the above mentioned deflector has been widely used in domestic drilling practice.

In addition, other methods can be used to locate the deflector, including the use of self-directional instruments, which are introduced into the drilling pipe. The working principle of these instruments is the plumb line effect generated by the tilting device in a tilted position.

【Words and phrases】

I.

screw bottom hole engine	螺杆井底发动机
bottom of the oil well	工作面；［钻］油井底
spatial	adj. 空间的
helix	n. 螺线；螺旋线

II.

square	adj. 方的；正方形的
inclinometer	n. 倾角仪
elevator	n. 提升机；起重机；［钻］吊卡
reaction torque	翻转力矩
compass	n. 罗盘；指南针
guidance	n. 指导
self-directional	自我定向

Course practice

1. Complex sentence analysis.

（1）A directional well drilling refers to a well that specifically points to a point that is "vertically projected away from its mouth".

refers to...：指的是……表示某个词或短语是对另一个事物的描述或提及。

【译文】定向钻井是指特定指向"远离井口的垂直投影"点的井。

（2）These engines have higher power than turbo drills and lower shaft speeds, which are beneficial for curvature setting.

higher than...：高于……表示一个数值、程度或位置比另一个数值、程度或位置更高。

【译文】这些发动机具有比涡轮钻机更高的功率和更低的轴速，这有利于曲率设置。

（3）It consists of three parts: the first part is vertical, the second part is the float curve, and the third part is the float line.

consists of...：包含……；由……组成；充斥着……。

【译文】它由三部分组成：第一部分是垂直的，第二部分是浮动曲线，第三部分是浮子线。

（4）This type of oil well is drilled in areas where geological conditions have a significant impact on the spontaneous bending of the wellbore.

have an impact on...：对……产生影响，表示某事物对另一事物产生重要或显著的影响。

【译文】这种类型的油井是在地质条件对井筒自发弯曲有重大影响的地区钻探的。

（5）Then take out the inclinometer from the drill pipe...

take out of/from...：从……中取出。从某个地方或物体中移除或提取某物……

【译文】然后从钻杆上取出倾角仪。

2．Translate the following paragraphs.

（1）Section Ⅲ［Fig. 3-29（c）］is smaller than the first two. It consists of two parts, the first part is vertical, and the second part is drawn along a curve that gradually increases the inclination angle of the barrel.

（2）When it is necessary to maintain the predetermined angle of the wellbore entering the formation, drilling can be carried out according to this profile.

3．Choose the proper answer to fill in the blanks and translate the sentences.

(in, of, up, from, on, with, at, between)

（1）Directional drilling is currently widely used（　　）oil, natural gas, and solid mineral drilling.

（2）The rotor represents the intermittent process（　　）continuous cutting (lateral) to bend the wellbore.

（3）The fourth part is made（　　）of the falling curvature curve.

（4）This profile is different（　　）previous profiles in that a curve segment 4 is added to vertical profile 1, profile 2.

（5）The above profile is a curve located（　　）the same vertical plane.

（6）The y mark on the paper tape is merged（　　）the y mark on the УБТ coupling.

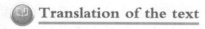
Translation of the text

钻井平台及其升降装置

Ⅰ．定向钻井

定向钻井是指特定指向"远离井口的垂直投影"点的井。定向钻井目前广泛应用于石油、天然气和固体矿产钻井（图3-28）。

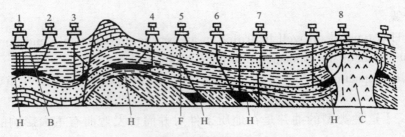

图 3-28 定向钻井示例

1—水下钻井；2—海上油田从海岸喷发；3—井筒从排出带（裂缝带）向含油带方向偏移；
4—当工作面位于无法安装钻机区域下方时，钻一个定向钻井；5—钻单斜油层；
6—钻辅助斜井以扑灭火灾或覆盖井喷；7—事故时偏航；8—在盐丘封闭区域钻定向钻井；
H—石油；B—水；F—气体；C—盐

钻孔方法有以下两种。

转子钻井 转子钻井即间歇地钻削（侧向）使井筒发生弯曲的过程。一般采用井底钻机，确保井筒的连续弯曲。

旋转钻井 在俄罗斯，绝大多数定向钻井是使用井底发动机完成的，而在其他国家，以旋转钻井为主，井底发动机主要采用规定方向的曲率组螺杆式井底发动机。国内外专家认为，给定方向的曲率组螺杆式井底发动机最有前途。这些发动机具有比涡轮钻机更高的功率和更低的轴速，这有利于曲率设置。

定向井的剖面

定向井的剖面应尽量减少钻井所需的资金和时间，确保完成钻井任务。

在定向钻井中，四种剖面分布最为广泛，如图 3-29 所示。

第一节［图 3-29（a）］是最常见的。它由三部分组成：第一部分是垂直的，第二部分是浮动曲线，第三部分是浮动线。这种剖面主要用于钻探 SCN 平均沉积深度偏差较大的单层矿床。

第二节［图 3-29（b）］由四个部分组成：垂直的第二部分上部由第三部分的上升曲率曲线组成——定向直线，第四部分由下降的曲率曲线组成。通常，该配置文件以略微修改的形式使用——没有区域 3，即曲率增加的区域 2 之后是曲率减少的区域 4。这种类型的剖面通常用于钻探深度达 2 500 m 的定向钻井。

第三节［图 3-29（c）］比前两节小。它由两部分组成：第一部分是垂直的，第二部分是沿着逐渐增加筒体倾角的曲线绘制的。当需要保持井筒进入地层的预定角度时，可以根据该剖面进行钻探。

第四节［图 3-29（d）］用于深层定向钻井。该轮廓与其他轮廓的不同之处在于，曲线段 4 被添加到垂直轮廓 1、沿着曲线绘制的轮廓 2 以及表示定向直线的轮廓 3 上。其特征是获得的曲率减小，这意味着筒体被剥离到垂直轮廓上，而直线轮廓 5 被添加到垂直轮廓上。当油井底部有多个生产层时，应使用这种类型的剖面。

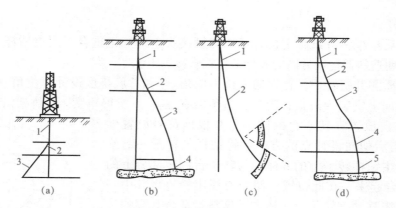

图 3-29 定向井轴剖面
(a) 第一节；(b) 第二节；(c) 第三节；(d) 第四节
1—垂直面积；2—曲率增大的部分；3—方向直线；4—曲线下降段；5—纵断面

上述轮廓是位于同一垂直平面上的曲线。这些配置文件称为常规配置文件。在钻井过程中，有时需要使用剖面，剖面类似于螺旋线或轴向螺旋的空间曲线，即空间剖面。这种类型的油井是在地质条件对井筒的自发弯曲有重大影响的地区钻探的。在构造这些油井的剖面时，应尽可能利用油井的自发弯曲模式，以最大限度地减少偏移钻井间隔。在俄罗斯，空间剖面定向钻井在格罗兹尼石油地区很常见。

Ⅱ. 斜井钻井

准备工作完成后，开始组装偏转布局。根据钻井工作计划组装布局，包括钻头、井底发动机、偏转装置（通常为曲线转换器），然后下入井中进行钻井。为了知道导流板的位置，在放置仪器时，将固定每个连接管上的标记位置。标签之间的距离由金属卷尺或其他最常见的方法确定——使用纸带，这是一种宽度为 8～10 cm、长度等于或略大于钻杆锁圈长度的厚纸条。将纸沿中线对折，并在其长度中间标记 o（偏转器）。

在曲线转换器上添加一个 o 标记，在纸带上的接头上添加一个 o 标记，并标记 y（图 3-30）。连接后，纸带上的 y 标记将与钻铤联轴器上的标记合并。在纸带上钻杆接头上的标记处画上标记 1，并将钻铤-钻杆组合体沿钻杆长度方向下放入井中。

拧紧并固定第二根管道，将纸带上的标记 1 与第一根管道的离合器上的标记结合，将第二根管的接头上的标记 2 与纸带结合，并划掉前一根管道。因此，没有纸带记录所有下降钻杆标签之间的距离。所有钻杆下放后，

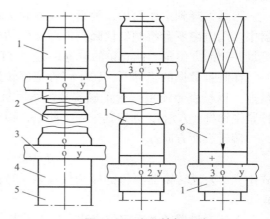

图 3-30 定向钻杆下放
1—钻杆；2—钻铤；3—纸带；4—转换器；
5—发动机底座；6—主钻杆

拧紧主钻杆。

将最后一根钻杆纸带上的标记 3 与管接头上的标记重合，用主钻杆（方钻杆）将用于指明造斜器运动方向的 o 标记打在转接器上。

在地质测井资料中，存在输入桥方向角 $\varphi_\text{м}$ 和井底挠度设计方位角 $\varphi_{\text{пр}}$。为了在所需方向上安装导流板，定义了差值 $\lambda=\varphi_{\text{пр}}-\varphi_\text{м}$。所得角度 λ 从桥架沿顺时针或逆时针方向放置在转子台的圆上，并根据符号放置标记 Π 以指示朝向设计点的方向。将钻杆与转子一起旋转，并将主钻杆转子上的 o 标记与转子台固定部分上的 o 标记结合起来。在此位置，使用模板用粉笔将其中一个方形边缘移动到转子上。通常，这是最适合观察的边缘。将工具放在适当的位置，取下升降器，然后将其冲洗至工作表面。到达工作面后，导流板应指向指定的方向，主钻杆转子台上的 o 标记应与转子上的 o 标记相同（图 3-31）。

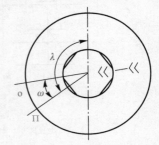

图 3-31　转子偏转方向

在提升过程中，钻杆按照与钻孔过程相同的顺序安装在底座上。在钻井过程中，根据钻井发动机的反作用扭矩和钻杆的长度，在新下放的管道上填充标记，并按照上述方式定位钻杆的偏转器位置 ω。通常，在实际工作中，168 mm 和 140 mm 钻杆每 100 m 的螺纹角 ω 分别为 3° 和 5°（钻杆长度不超过 1 000～1 500 m）。

偏转方向

钻杆可以下放到工作面坡度大于 3° 的斜井中，就像钻普通的立井一样。在这种情况下，使用电磁罗盘和带磁转换器的倾角仪将工作面偏转器指向正确的方向。用磁罗盘和磁转换器倾角仪偏转器的钻孔定位是基于多点倾角仪的，使用电磁罗盘测量方位。该工具包括一根钻杆柱，其末端为底部的抗磁性管（1×18N9T 钢或 D16 T 铝合金）。拧紧反磁管底部的转换器，以固定其中的磁场源。磁转换器连接到极限转换器和曲线转换器，后者之间交叉连接。接下来是发动机。

工具下降到工作面后，拧紧转换器和旋转套筒，并将伐木滚筒固定在上管离合器中。用电磁罗盘将倾斜仪降到电缆上的钻杆内。在磁转换器上方的反磁管中预先测量井筒的方位角。偏心轮的位置是通过磁转换器中的点采样来确定的。

在测量数据的指导下，转子将管道旋转到所需的导流板位置，然后重新校准测量点，以检查导流板是否正确安装。然后从钻杆上取出倾角仪，标记仪器的位置，拧紧主钻杆（方形），固定其中一个肋条，锁定转子，开始钻孔。

偏转表面方向：

$$\beta=360°-\varDelta+\varphi$$

式中，$\varDelta$ 为磁转换器测得的倾角仪读数（°）；φ 为在反磁管中测量时，井筒的方位角（°）。

上述导流器的井底定位方法在国内钻井实践中得到了广泛应用。

此外，还可以使用其他方法来定位导流板，包括使用引入钻杆的自定向仪器。这些仪器的工作原理是由倾斜位置的倾斜装置产生的铅垂线效应。

Evaluate

任务名称		钻井平台及其升降装置	姓名	组别	班级	学号	日期	
		考核内容及评分标准		分值	自评	组评	师评	平均分
三维目标	知识	了解定向钻井、斜井钻井装置及钻井方法等		25分				
	技能	能阅读专业英语文章、翻译专用英语词汇		40分				
	素养	在学习过程中秉承科学精神、合作精神		35分				
加分项	收获（10分）	收获（借鉴、教训、改进等）：	你进步了吗？				加分	
			你帮助他人进步了吗？					
	问题（10分）	发现问题、分析问题、解决方法、创新之处等：					加分	
总结与反思							总分	

Module 4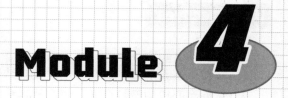

Typical Electromechanical Equipment

Focus

- Unit 1   Construction Equipment
- Unit 2   Medical Apparatus and Instruments
- Unit 3   Other Equipment

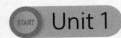 **Unit 1　Construction Equipment**

Unit objectives

【Knowledge goals】
1. 掌握供浆泵的型号意义、用途及其结构等。
2. 掌握离心机的型号意义、用途及其结构等。

【Skill goals】
1. 能够阅读专业英语文章。
2. 可以翻译专用英语词汇。

【Quality goals】
1. 在学习过程中发扬科学研究精神。
2. 增强团队合作意识。

1.1　Slush Supply Pump

建筑工程设备

General Description

Model GJ slush supply pump is a vertical single-suction submerged pump suitable for transporting mud, ore slush, sewage, and other similar liquid in oil field, metallurgy, construction, environmental protection and other industries. The suction inlet is on upper pump, with vane suction inlet upward, so there is no need to use any shaft seal and the frequent faults of former submersed pump's shaft seal are also solved. The advantage of this structure is no shaft seal (such as filler seal, mechanical seal, etc.) so power loss (shaft seal friction force) is reduced and stopping fault caused by changing and maintaining shaft seal parts is also avoided. The axial force generated by the pump during operation is upward, balances the rotor mass (partly or wholly), which improves the operation condition for upper bearing, extends the life span of bearing. The structure of this model pump is a national patent.

The model meaning of supply pump is shown in Fig. 4-1.

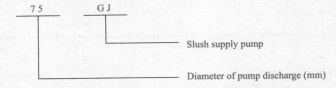

Fig. 4-1　The model meaning of slurry pump

The main technical parameters of model GJ slush supply pump is shown in Table 4-1.

Table 4-1 Main technical parameters of model GJ slush supply pump

Model	Flow/ (m³·h⁻¹)	Lift/m	Motor power/kW
65GJ	40	15	5.5
75GJ	50	15	7.5
75GJ	60	14	11

Structure Introduction

The model pump is vertical single-stage centrifugal pump, consisting of casing, vane, shaft, pump cover, bearing body, linking pipeline, support, motor, etc. Casing, pump cover, bearing body, linking pipeline, support compose flow-conducting parts and support parts, vane and shaft compose rotor parts, which are supported in support parts by upper and lower bearings. The vertical motor is fixed on the top joined by the claw shape coupling and pump shaft. Between two bearings there is refueling oil cup or oil hole. Fig. 4-2 shows the overall structure of the centrifugal pump.

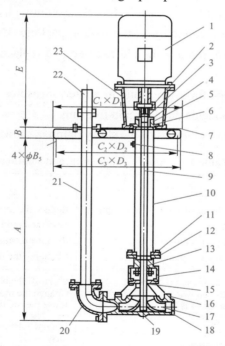

Fig. 4-2 The overall structure of the centrifugal pump

1—Explosion insulation motor; 2—Elastic block; 3—Oil cup; 4—Bearing; 5—Bearing body II; 6—Oil seal; 7—Pump seat; 8—Oil filler plug; 9—Shaft; 10—Connecting hose; 13—Filler; 14—Bearing body I; 15—Sliding bearing; 16—Pump cover; 17—Vane; 18—Casing; 19—Vane nut; 20—Elbow; 21—Discharge pipeline; 22—Hose connector; 23—Motor support

Working Principles and Features

When the pump is working, the motor revolves pump shaft through shaft coupling.

Liquid is drawn into vane though upper casing, enters casing after getting energy, and drains from pump discharge.

Because structure of this model pump is special, the pump can operate normally without leaking though there is no shaft seal. The trouble caused by renewing and maintaining shaft seal is avoided and shutdown time for repairing is reduced. The axial force generated by the pump during operation can balance the rotor mass, which improves operation condition for bearing, reduces power loss (friction force), and heightens the pump efficiency accordingly.

Operation and Maintenance

(1) Check if all tight parts are tightened before operating pump.

(2) Fill lithium□lubricating grease into the upper bearing, machine oil No.20 into the lower one.

(3) Check if the motor rotation is correct. This type of pump rotates counterclockwise from the motor direction.

(4) During pump use, the bearing temperature should be checked, and the maximum temperature should not exceed 80 ℃.

(5) Add lubricating oil (grease) every 10–15 days during use.

(6) Check the sealing condition of the sealing packing. If the packing gland is too loose and properly pressed, remove the old packing and replace it with a new one every month of operation.

(7) If the pump stops for a long time in winter, measures should be taken to avoid damaging the pump components due to liquid freezing.

The possible faults and solutions of the slurry pump is shown in Table 4-2.

Table 4-2 Slurry pump faults and solutions

Faults	Reason	Solution
Low or no flow rate	The liquid level drops below the pump suction port; The suction port is blocked; Reverse rotation of pump	Reduce the installation height of the pump or raise the liquid level; Remove blockages; Changing the direction of motor rotation
Lower lift	The flow duct of vane is blocked; Pump rotates counter; The rotating speed is low	Clear off the blocking things; Change the motor direction; Raise the rotating speed
Excessive temperature of bearings	Lubricant or grease has impurities; Bearing lubricating is not good	Change lubricant or grease after cleaning the dirty ones; Replenish lubricant or grease
Overloading of motor	Flow is higher; Medium gravity is bigger	Reduce flow; Reduce flow or medium gravity
Violent vibration or loud noise	Bearing is damaged; Vane nut goes loose; Shafts of pump and motor are not concentric; Vane is heavily worn	Change bearing; Tighten the vane nut; Calibrate to be concentric; Change vane

【 Words and phrases 】

submerge	v. （使）潜入水中；（使）没入水中；浸没；湮没
submerged pump	潜水泵
mud	n. 泥；淤泥；泥浆
slush	n. 烂泥；污水；水泥砂浆
sewage	n. 污水；污物
metallurgy	n. 冶金；冶金学
suction	n. 吸；吸力；抽吸
vane	n. 叶片；［气象］风向标；风信旗
shaft seal	轴封：一种用于密封旋转轴与固定部件之间的间隙的机械设备
friction force	摩擦力
shaft coupling	联轴器（一种用于连接两个轴的机械部件）
casing	n. 套；盒
lithium	n. 锂（符号 Li）
grease	n. 油脂；润滑脂
	v. 给……加润滑脂；为……涂（或抹）脂
lithium-lubricating grease	锂基润滑脂
elastic block	弹性块
sliding bearing	滑动轴承（一种用于支承旋转或滑动部件的机械元件）

Course practice

1. Complex sentence analysis.

（1）Model GJ slush supply pump is a vertical single-suction submerged pump suitable for transporting mud, ore slush, sewage...

submerge：（使）潜入水中，（使）没入水中，浸没；湮没。

suitable for...：适合……的，适合于……

【译文】GJ 型供浆泵是一种立式单吸潜水泵，适用于输送泥浆、矿泥、污水……

（2）The model pump is vertical single-stage centrifugal pump, consisting of casing, vane, shaft, pump cover, bearing body...

consist of...：由……组成，表示某个整体是由若干部分组成的。

【译文】该型号泵为立式单级离心泵，由泵体、叶轮、轴、泵盖、轴承体……组成。

（3）The vertical motor is fixed on the top joined by the claw shape coupling and pump shaft.

be fixed on...：根据，以……为基础；建立在……基础上。

【译文】立式电机通过爪形联轴器与泵轴连接固定在顶部。

（4）Liquid is drawn into vane though upper casing, enters casing after getting energy, and drains from pump discharge.

be drawn into：被卷入，指无意中或不情愿地被卷入某种情况、争论或纷争中。

【译文】液体从泵体上侧被吸入叶轮，通过叶轮获得能量后进入泵体，从泵排出口排出。

2．Translate the following sentences.

（1）The model pump is vertical single-stage centrifugal pump, consisting of casing, vane, shaft, pump cover, bearing body, linking pipeline, support, motor, etc.

（2）Casing, pump cover, bearing body, linking pipeline, support compose flow-conducting parts and support parts; vane and shaft compose rotor parts, which are supported in support parts by upper and lower bearings.

3．Choose the proper answer to fill in the blank and translate the sentences.

（through, by, to, at, in）

（1）When the pump is working, the motor revolves pump shaft（　　）shaft coupling.

（2）The trouble caused（　　）renewing and maintaining shaft seal is avoided and shutdown time for repairing is reduced.

 Translation of the text

供浆泵

概述

GJ 型供浆泵是一种立式单吸潜水泵，适用于输送泥浆、矿泥、污水以及与此相类似的液体。其吸入口在泵体上侧，叶轮吸入口向上，因此，不需要任何形式的轴封装置，解决了传统液下泵轴封装置经常失效的难题。该结构的优点为无轴封装置（如填料密封、机械密封等），因而减少了动力损失（轴封摩擦力），使用中避免因更换、维护轴封装置而造成停机故障。该结构的泵工作时转子部件产生的轴向力向上，能够平衡（全部平衡或部分平衡）转子质量，改善上部轴承的工作条件，延长了轴承使用寿命。该型泵结构为国家专利。

供浆泵型号及其含义如图 4-1 所示。

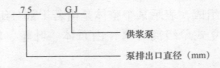

图 4-1　供浆泵型号及其含义

供浆泵主要性能参数如表 4-1 所示。

表 4-1 供浆泵的主要性能参数

泵型号	流量 /(m³·h⁻¹)	扬程 /m	电机功率 /kW
65GJ	40	15	5.5
75GJ	50	15	7.5
75GJ	60	14	11

结构说明

该型号泵为立式单级离心泵，由泵体、叶轮、轴、泵盖、轴承体、连接管、支架、电机等组成。泵体、泵盖、轴承体、连接管、支架组成导流和支承部件，叶轮、轴组成转子部件，转子部件由上下两轴承支承于支承部件中。立式电机通过爪形联轴器与泵轴连接固定在顶部，在两轴承处设有加油油杯或油孔。图 4-2 所示为离心泵的整体结构。

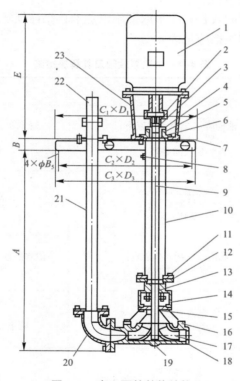

图 4-2 离心泵的整体结构

1—防爆绝缘电机；2—弹性块；3—油杯；4—轴承；5—轴承体Ⅱ；6，12—油封；7—泵座；8—注油塞；9—轴；10—连接管；13—填料；14—轴承体Ⅰ；15—滑动轴承；16—泵盖；17—叶轮；18—泵体；19—叶轮螺母；20—弯头；21—排出管；22—胶管接头；23—电机支架

工作原理和特点

泵工作时，电机通过联轴器带动泵轴旋转。液体从泵体上侧被吸入叶轮，通过叶轮获得能量后进入泵体，从泵排出口排出。

由于该型泵的结构特殊，泵无须轴封装置即能保证泵在无泄漏的情况下正常工作，避免更换、维修轴封装置带来的麻烦并减少了维修停机时间。泵工作时产生的轴向力能平衡转子质量，大幅改善了轴承工作条件，减少了动力损失（摩擦力），使泵效率相应提高。

使用、维护和保养

（1）使用前检查所有紧固件是否牢固。

（2）上部轴承注入锂基润滑油（脂），下部轴承注入20号机油。

（3）检查电机转向是否正确，该型号泵从电机方向看电机为逆时针方向旋转。

（4）泵使用中应检查轴承温度，最高温度不得超过80℃。

（5）使用中每10～15天加注一次润滑油（脂）。

（6）检查密封填料的密封情况。如太松，适当压紧填料压盖，每运行一个月，应将旧填料取出换上新填料。

（7）冬季若停泵时间较长，应采取措施以免液体结冰损坏泵件。

供浆泵故障及其解决措施见表4-2。

表4-2 供浆泵故障及其解决措施

故障	原因	解决方法
流量偏低或无流量	液面降至泵吸入口以下；吸入口被堵塞；泵反向旋转	降低泵安装高度或升高液面；清除堵塞物；改变电机转向
扬程偏低	叶轮流道堵塞；泵反向旋转；泵转速低	清理堵塞物；改变电机转向；提高泵转速
轴承温度过高	润滑油（脂）有杂质；轴承润滑不良	清理脏油（脂），更换新油（脂）；加注润滑油（脂）
电机过载	泵流量过大；介质重力过大	调小流量；调小流量或减小介质重力
剧烈振动或噪声大	轴承损坏；叶轮螺母松动；泵轴与电机轴不同心，叶轮磨损严重	更换轴承；紧固叶轮螺母；调整至同心；更换叶轮

Evaluate

任务名称		供浆泵		姓名	组别	班级	学号	日期	
考核内容及评分标准				分值	自评	组评	师评	平均分	
三维目标	知识	了解供浆泵型号的意义、用途及其结构等		25 分					
	技能	能阅读专业英语文章、翻译专用英语词汇		40 分					
	素养	在学习过程中秉承科学精神、合作精神		35 分					
加分项	收获（10 分）	收获（借鉴、教训、改进等）：		你进步了吗？			加分		
				你帮助他人进步了吗？					
	问题（10 分）	发现问题、分析问题、解决方法、创新之处等：					加分		
总结与反思							总分		

1.2 Centrifuge

 Text

General Description

Type of LW400-NY-G and LW500-NY centrifuges (brief for the "two-type centrifuges") are special devices to separate solid phase from well drilling liquid, which are new generation type developed and designed according to the actual demands of oil drilling and on the base of studying current domestic and foreign centrifuges. The maxmum disposal of type LW500-NY may reach 60 m^3/h which make the well drilling liquid higher quality and shorten the time of liquid disposal. Without disassembling the belt wheel, type of LW400-NY-G centrifuge can complete the transformation from slow speed to high speed only using one belt wheel, i.e., the drum can complete two rotating speed, 1,800 r/min and 2,400 r/min. This solve, the problem of choosing different centrifuge to match different well drilling mine shaft. Hard alloy is inlayed in the conveyor vanes of the two-type centrifuges lengthening their service time several times.

Application

The two-type centrifuges mainly used to control the density and viscosity, to reduce the solid content of well drilling liquid, to remove the harmful solid phase or recover barite, and to recover and reuse the bottom flow so as to guarantee the high performance of well drilling liquid and raise drilling speed.

After the well drilling liquid is processed, not only the density and the viscosity are controlled, but also its sand content is cut down greatly. Therefore, the machine service time is extended. The economic benefit and social benefit are obviously obtained.

Because the two-type centrifuges have features of reasonable design, good adaptability, high processing volume, easy operation and convenience of maintenance, and timely supply of parts, they are very popular with customers.

Model Meaning and Main Technical Parameters

1. Description of Machine Type (Fig.4-3)

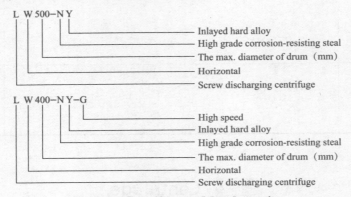

Fig.4-3 Centrifuge model and meaning

2. Main Parameters (Table 4-4)

Table 4-4 Technical parameters of the centrifuges

Parameters and the units	LW400-NY-G		LW500-NY
Drum speed / (r · min^{-1})	1,800	2,400	1,800
Max.processed voiume / (m^3 · h^{-1})	40	30	60
Main motor power /kW	30	30	37
Auxiliary motor power /kW	5.5	5.5	7.5
Separation factor	689	1,224	906
Total mass/ kg	2,800		3,500

Construction and Working Principle

1. Construction

The two-type centrifuges consist of the main motor, coupler, differential, drum assembly, upper cover of plough, body of plough, auxiliary motor, isolated anti-exploration control unit, general base and other auxiliary equipment (Fig. 4-5).

Drum assembly consists of straight drum, corn drum, flange shaft with big and small end, conveyor, and other parts. Straight drum, corn drum and conveyor are concentrically fixed on both end bearing base of plough through flange shaft. The spline of the differential is connected with the spline jacket of the conveyor through spline shaft. The shell is driven by the flange shaft which connects the big flange and drum assembly. Various speed can be obtained through adjusting the rotating speed of input shaft of the differential. There are 6 adjustable overflow plates on the flange of big flange shaft. The subsidence area will be controlled through changing the covering area of the damper one overflow hole, thus the separating result will be changed.

The conveyor consists of column and corn drum body, closure, rail splice, screw vanes welding assembly, flange shaft, spline and spline shaft. The screw vanes are evenly welded on the drum body. The vanes are inlayed hard alloy, so that wearability of vanes is raised and the service time is extended. There are 26 holes (ϕ50 mm) on the corn drum (LW400-NY-G22, LW500-NY26). Wear-resisting alloy is sprayed on the circle of the holes. The spline jacket is connected with the spline on the column drum section.

The differential is involute epicycilc gear type. It consists of two stage epicyclic gears. The surface of sun and epicyclic gears are hard processed, so that they have higher ability of bearing load. One safe bolt is placed on the input shaft end. When the push torque of conveyor is overload, the bolt will be cut down and the drum will be synchronous running with the conveyor and thus the centrifugal will be protected. Before restarting, the bottom end cover should be open and replace new bolt.

Interlocking unit is adopted to control both main and auxiliary motors. When starting the machine, the auxiliary, motor should be started firstly and then the main motor will be started in 20 s. When stopping the machine, the main motor stops while pressing the button, the interlocking unit will automatically cut down the auxiliary motor in 30 s after the main motor stops. Meanwhile, a special anti-exploration breaker is provided slush-supplying pump after starting the main motor.

2. Working Principle

The well drilling liquid to be processed is sent to the input pipe of the drum small end by slush supplying pump and then to the liquid chamber, finally into drum through the suction of conveyor. The solid phase of well drilling liquid, on the action of centrifugal force produced by the drum rotation at a high speed, is thrown to the drum inner wall and subside down. Meanwhile, the conveyor, driven by the output spline of differential, rotates as the same direction as the drum, however conveyor speed is a little slower than the drum, the speed difference is formed between the drum and the conveyor. The conveyor scrapes subsided slag on the inner wall of the drum and pushes them to the dry zone at the small end of the drum. The slag is drained from the slag outlet to slag land. The separated well liquid will overflow from the overflow port at the big end of drum and them drain out through the discharging hole on the plough, the separation has realized.

Parameters of Attached Device

1. Slush Supplying Pump

(1) Main parameters of slush supplying pump (Table 4-5).

Table 4-5　Main parameters of slush supplying pump

Types of centrifuges	LW400-NY-G	LW50G-NY
Type of slush supplying pump matched	65GJ	75GJ
Power of slush supplying pump /kW	5.5	11
Speed of slush supplying pump / $(r \cdot min^{-1})$	1,440	1,460
Lift of slush supplying pump / m^{-1}	15	14
Capacity of slush supplying pump / $(m^3 \cdot h^{-1})$	40	60

(2) Performance. Slush supplying pump supplies slush for the centrifuge, which is desirable device to deliver fluid containing solid fiber impurity slush. It has features of high efficiency, energy saving, large exhaust, low noises small vibration, stable performance, etc.

2. Hydraulic Coupler

(1) Main technical parameters of the centrifuges (Table 4-6).

Table 4-6　Main technical parameters of the centrifuges

Types of centrifuges	LW400-NY-G	LW500-NY
Type of Hydraulic coupler	YOXp360	YOXp400
Delivery power /kW	30	37
Oil filled volume	5.8	10
Input rotating speed / $(r \cdot min^{-1})$	1,470	1,480

(2) Performance. It is a power delivery device by using hydraulic delivery, which has features of steady start with load. It can improve motors' starting performance and

protect motor when overload, also isolate twisting vibration and impact.

(3) Notes in operation.

① The working fluid temperature is not over 90 ℃.

② It is not allowed to open the eutectic plug when coupler temperature is high, so as not to occur scald accident.

③ It is not allowed to change the working liquid medium of coupler randomly. Any filling should be done strictly in accordance with specific regulation.

④ Do not disassemble the coupler randomly so as not to break sealing ring and balance precision.

⑤ To check the working oil quality after 3,000 of running. The oil should be changed if it is found bad.

【Words and phrases】

centrifuge	n. 离心机
	v. 用离心机分离；使……受离心作用
disposal	n. 处理；清除；（土地、财产等的）变卖；转让
disassembling	n. 拆卸；反汇编
	v. 拆开；分散（disassemble 的 ing 形式）
belt wheel	皮带轮
transformation	n. (彻底或重大的) 改观；变化；转变
viscosity	n. 黏性；黏度
barite	n. [矿物] 重晶石
convenience	n. 方便；便利；便利的事物；便利设施；公共厕所
screw	n. 螺丝（钉）；螺杆；螺旋桨
screw discharging centrifuge	螺旋卸料离心机
auxiliary	adj. 辅助的；备用的；后备的
	n. 助手；辅助人员
differential	adj. 微分的；差别的；特异的
	n. 微分；差别
drum assembly	[机] 摩擦鼓部件
plough	v. 耕，犁；（用或仿佛用犁）划出（沟或线）
flange	n. (机) 法兰；(机)(古生) 凸缘；轮缘；边缘
	vt. 给……装凸缘
rotating speed	穿引速度；引纸抄速
conveyor	n. 传送带；传送装置；传播者；传达者
accumulate	v. 积累；积攒

Course practice

1. Complex sentence analysis.

（1）The conveyor scrapes subsided slag on the inner wall of the drum and pushes them to the dry zone at the small end of the drum.

slag on：随便穿上；胡乱涂上；断然拒绝。

【译文】输送器将滚筒内壁上的沉渣刮下来，并将其推至滚筒小端的干燥区。

（2）The slag is drained from the slag outlet to slag land.

be drained from：被抽干；drain：排出，滤干，喝光。

【译文】沉渣从排渣口进入排渣槽。

2. Translate the following paragraphs.

（1）The conveyor consists of column and corn drum body, closure, rail splice, screw vanes welding assembly, flange shaft, spline and spline shaft. The screw vanes are evenly welded on the drum body. The vanes are inlayed hard alloy, so that wearability of vanes is raised and the using life is extended.

（2）There are 26 holes（φ50 mm）on the corn drum（LW400-NY-G22, LW500-NY26）. Wear-resisting alloy is sprayed on the circle of the holes. The spline jacket is connected with the spline on the column drum section.

3. Choose the proper answer to fill in the blank and translate the sentences.

（of, on, to, for, from）

（1）It consists（　　）two stage epicyclic gear.

（2）One safe bolt is placed（　　）the input shaft end.

（3）Interlocking unit is adopted（　　）control both main and auxiliary motors.

（4）The well drilling liquid to be processed is sent（　　）the input pipe of the drum small end.

（5）Slush supplying pump supplies slush（　　）the centrifuge.

Translation of the text

离心机

概述

LW400-NY-G 型离心机、LW500-NY 型离心机（以下简称"两型离心机"）是对钻井液进行固液分离的专用设备，它们是根据当前石油钻井实际需要，并参照当前国内外对离心机的研究而研制的新一代固液分离设备。LW500-NY 型离心机最大处理量可达 60 m³/h，提高了钻井泥浆质量，且缩短了泥浆处理时间。LW400-NY-G 型离心机在不拆卸皮带轮的情况下，用同一根皮带就能实现从低速到高速的转变，即滚筒能实现 1 800 r/min 和 2 400 r/min 两种转速，从根本上解决了因钻井工矿不同而另选离心机进行匹配的问题。两型离心机在输送器叶片上均镶嵌了硬质合金块，使用寿命可提高几倍。

用途

两型离心机主要用于控制钻井液的密度、黏度，降低钻井液的固相含量，清除有害固相或回收重晶石，以及对旋流器底流进行二次回收利用以保证钻井液的优良性能，提高钻井速度。

经离心机处理过的钻井液，不但控制了密度、黏度，而且使其含沙量大幅降低，提高了设备的使用寿命，可以取得明显的经济效益和社会效益。

由于两型离心机具有设计合理、适应性强、处理量大、操作简单、维护方便，且配件供应及时的特点，所以很受用户欢迎。

型号含义、主要技术参数

1. 型号含义（图4-3）

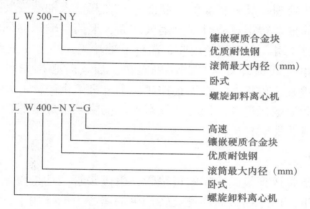

图4-3 离心机型号含义

2. 主要技术参数（表4-4）

表4-4 离心机技术参数

技术参数及单位	LW400-NY-G		LW500-NY
滚筒转速 / (r·min^{-1})	1 800	2 400	1 800
最大处理量 / (m^3·h^{-1})	40	30	60
主电机功率 / kW	30	30	37
副电机功率 / kW	5.5	5.5	7.5
分离因数	689	1 224	906
总质量 / kg	2 800		3 500

结构及工作原理

1. 结构

两型离心机主要由主电机、耦合器、差速器、滚筒轴总成、犁上盖、犁底座、犁箱体、副电机、隔爆型离心机控制装置和大底座及其他附属设备组成（图4-5）。

滚筒轴总成主要由直滚筒、锥滚筒、大小端法兰轴和输送器等部件组成。直滚筒通过两端大小法兰轴与输送器同心安装在犁两端的轴承座上。差速器输出花键盘，通过花键轴与输送器的花键套连接。外壳通过连接法兰与滚筒轴总成的大端法兰轴连接传动。通过调整差速器输入轴的转速可得到不同的差速。大端法兰轴的法兰上有6个可调节的溢流板，改变挡板对溢流孔遮挡面积，就可控制沉降区的大小，改变其分离效果。

输送器由柱锥形筒体、隔板、连接板、螺旋叶片、内尾法兰轴、花键套和花键轴组成。螺旋叶片均匀地焊接在筒体上。叶片表面镶嵌硬质合金块，增加了叶片的耐磨性和使用寿命。筒体锥直段上开有26个$\phi50$圆孔（LW400-NY-G型离心机有22个，LW500-NY型离心机有26个）。孔周围喷焊有耐磨合金。花键套在筒体柱段与花键轴连接。

差速器为渐开线行星齿轮差速器，它由两级行星齿轮传动组成。太阳轮、行星轮均进行表面硬化处理，具有较高的承载能力并在其输入轴端安装扭矩过载保护安全销。当输送器的推料转矩过载时安全销被切断，滚筒与输送器同步旋转，从而起到了保护离心机的作用。在重新使用前，需拆下轴端盖，换上安全销。

电控箱采用电气联锁装置来控制主副电机。启动时必须先启动副电机，大约在20 s后启动主电机。停机时，按停止按钮，主电机停止运行，大约30 s后，电器联锁装置自动切断副电机电源。同时还设有专门的供浆防爆断路器，在主电机启动后，能方便地控制供浆泵的开启。

2．工作原理

钻井液由供浆泵送到滚筒小端的进浆管，由进浆管到达进液室，最后通过输送机的进浆孔进入滚筒内。在滚筒高速旋转产生的离心力的作用下，钻井液中固相被甩到滚筒内壁上沉淀下来。同时输送机在差速器输出花键轴的驱动下，与滚筒同方向旋转，但转速比滚筒稍慢一些，形成滚筒与输送机之间的差速。输送器将滚筒内壁上的沉渣刮下来，并将其推到滚筒小端的干燥区。沉渣从排渣口进入排渣槽。分离后的钻井液，从滚筒大端溢流口溢出，经犁上的排液孔排出，从而实现分离的目的。

连接设备的性能参数

1．供浆泵

（1）供浆泵主要技术参数（表4-5）。

表4-5　供浆泵主要技术参数

所属离心机型号	LW400-NY-G	LW500-NY
配用供浆泵型号	65GJ	75GJ
供浆泵功率 / kW	5.5	11
供浆泵转速 / (r·min^{-1})	1 440	1 460
供浆泵扬程 / m	15	14
供浆泵流量 / (m^3·h^{-1})	40	60

（2）性能。供浆泵给离心机提供泥浆，是输送含有固体纤维杂质浆料流体的理想设备。它具有高效、节能、大排量、噪声低、振动小、性能稳定等优点。

2．液力耦合器

（1）主要技术参数（表 4-6）。

表 4-6　液力耦合器的技术参数

离心机型号	LW400-NY-G	LW500-NY
液力耦合器型号	YOXp360	YOXp400
传递功率 / kW	30	37
充油量 / L	5.8	10
输入转速 / (r·min^{-1})	1 470	1 480

（2）性能。它是利用液力传递动力的装置，具有负载时启动平稳的特征。它能改善电机的启动性能和对电机的过载起保护作用，能隔离扭振和冲击等。

（3）使用时的注意事项。

①连续运转时，工作液温度不得超过 90 ℃。

②耦合器温度较高时，严禁打开易熔塞或易爆塞，以防发生烫伤事故。

③不允许随意改变耦合器工作液介质，应严格按规定充液。

④不要随意拆卸耦合器，以免密封环和平衡精度被损坏。

⑤运转 3 000 h 后应检查工作油品质。如发现油质变坏，应予以更换。

Evaluate

任务名称		离心机		姓名	组别	班级	学号	日期
		考核内容及评分标准		分值	自评	组评	师评	平均分
三维目标	知识	了解离心机的型号意义、用途及其结构等		25 分				
	技能	能阅读专业英语文章、翻译专用英语词汇		40 分				
	素养	在学习过程中秉承科学精神、合作精神		35 分				
加分项	收获（10 分）	收获（借鉴、教训、改进等）：		你进步了吗？			加分	
				你帮助他人进步了吗？				
	问题（10 分）	发现问题、分析问题、解决方法、创新之处等：					加分	
总结与反思							总分	

Unit 2 Medical Apparatus and Instruments

Unit objectives

【Knowledge goals】
1. 掌握医用监护仪的概念及用途等知识。
2. 掌握除颤的概念，除颤仪的定义、结构及分类等相关知识。

【Skill goals】
1. 能够阅读专业英语文章。
2. 可以翻译专用英语词汇。

【Quality goals】
1. 在学习过程中发扬科学研究精神。
2. 增强团队合作意识。

2.1 Monitor

Text

Ⅰ. Overview of Medical Monitors

The Concept of a Monitor

Medical monitoring instruments can continuously monitor the physiological parameters of the human body for a long time, and can store, display, analyze, and control the test results. When abnormal situations occur, alarms are issued to remind medical staff to handle them in a timely manner (Fig. 4-4).

Fig. 4-4 Schematic diagram of the monitor

The Purpose of the Monitor

This patient monitor is intended to be used for monitoring, displaying, reviewing, storing and transferring of multiple physiological parameters including electrocardiogram (ECG), heart rate (HR), respiration (Resp), temperature (Temp), SpO_2, pulse rate (PR), non-invasive blood pressure (NIBP), invasive blood pressure (IBP), cardiac output (CO), carbon dioxide (CO_2), oxygen (O_2), anesthetic gas (AG), impedance cardiograph (ICG), bispectral index (BIS) and respiration mechanics (RM) of single adult, pediatric and neonatal patients.

Interpretation of resting 12-lead ECG and CO monitoring are restricted to adult patients only. The ICG is only for use on adult patients who meet the following requirements-height: 122–229 cm, mass: 30–159 kg. The arrhythmia detection, ECG ST-segment, BIS and RM monitoring are not intended for neonatal patients.

This monitor is to be used in healthcare facilities by clinical professionals or under their direction. It is not intended for helicopter transport, hospital ambulance, or home use.

Introduction to the Display Screen of the Monitor

This patient monitor adopts a high-resolution TFT LCD to display patient parameters and waveforms. Fig.4-5 shows a typical monitor display interface.

1. Patient Information Area

This area shows the patient information such as department, bed number, patient name, patient category and paced status.

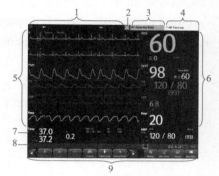

Fig. 4-5 Typical monitor display interface

1—Patient information area; 2—Alarm symbols;
3—Technical alarm area; 4—Physiological alarm area;
5—Waveform area; 6—Parameter area A;
7—Parameter area B; 8—Prompt message area;
9—QuickKeys area

: Indicates that no patient is admitted or the patient information is incomplete.

: Indicates that the patient has a pacer.

If no patient is admitted, selecting this area will enter the "Patient Setup" menu. If a patient has been admitted, selecting this area will enter the "Patient Demographics" menu.

2. Alarm Symbols

: Indicates alarms are paused.
: Indicates alarm sounds are paused.
: Indicates alarm sounds are turned off.
: Indicates the system is in alarm off status.

3. Technical Alarm Area

This area shows technical alarm messages and prompt messages. When multiple messages come, they will be displayed circularly. Select this area and the technical alarm list will be displayed.

4. Physiological Alarm Area

This area shows physiological alarm messages. When multiple alarms occur, they will be displayed circularly. Select this area and the physiological alarm list will be displayed.

5. Waveform Area

This area shows measurement waveforms. The waveform name is displayed at the left upper corner of the waveform. Select this area and the corresponding measurement setup menu will be displayed.

6. Parameter Area A

This area shows measurement parameters. Each monitored parameter has a parameter window and the parameter name is displayed at the upper left corner. The corresponding waveform of each parameter is displayed in the same row in the waveform area. Select this area and the corresponding measurement setup menu will be displayed.

7. Parameter Area B

For the parameters displayed in this area, their corresponding waveform are not displayed.

8. Prompt Message Area

This area shows the prompt messages, network status icons, battery status icons, date and time, etc.

- : Indicates patient monitor is connected to a wire network successfully.
- : Indicates the patient monitor has failed to connect a wire network.
- : Indicates the wireless function is working.
- : Indicates the wireless function is not working.
- : Indicates a CF storage card is inserted.
- : Indicates a secondary display or remote display is connected.
- : "Screen Setup" button.

9. QuickKeys Area

This area contains QuickKeys that give you fast access to functions.

Clinical Application of Medical Monitors

According to different clinical nursing objects and monitoring purposes, clinical monitors are used for the following nursing monitoring: monitoring during and after surgery; delivery monitoring and fetal monitoring; monitoring of critically ill patients; monitoring of patients during the recovery period; monitoring of patients undergoing treatment (kidney dialysis, hyperbaric oxygen chamber, radiation therapy, mental illness, etc.).

Various special monitors, also called monitoring systems, are used in departments and wards as needed, mainly including critical patient monitors, coronary heart disease monitors, delivery monitors, neonatal and preterm infant monitors, intracranial pressure monitors, anesthesia monitors, sleep monitors, etc.

Physiological and biochemical parameters monitored by the monitor are ECG monitoring, respiratory monitoring, non-invasive blood pressure monitoring, invasive blood pressure monitoring, airway carbon dioxide monitoring, airway oxygen monitoring, blood volume auxiliary monitoring, blood oxygen saturation/blood volume monitoring, temperature monitoring, pH value monitoring, blood gas monitoring, etc. In addition to the monitoring function, the medical monitor also has the function of disease diagnosis and treatment, as well as the rescue function, such as Holter monitor, blood pressure monitor, heart defibrillation monitor, etc.

Structure of Medical Monitors

Modern medical monitoring instruments mainly consist of four parts, namely vital sign measurement components, host and system, connection interfaces, and sensors and electrodes. The brief description of the functions of each part is as follows.

The vital sign measurement component mainly includes various measurement components for measuring human vital signs such as electrocardiogram, respiration, body temperature, blood pressure, pulse oxygen saturation, and end-breathing carbon dioxide.

The host and system mainly include the main control board, display, keyboard, recorder and operation software, and other external expansion equipment.

There is a requirement for bidirectional communication between the vital sign measurement component and the host interface, so it is necessary to set feature recognition for each vital sign measurement module to facilitate targeted operations by the host.

The connection forms of sensors, connecting cables, and vital sign measurement components are divided into hard connection and intelligent connection.

Characteristics of Medical Monitors

(1) The safety performance is in line with international standards.

(2) More powerful functions and superior performance.

(3) Specialized monitors are developing rapidly.

(4) Remote monitoring and home monitoring are becoming increasingly popular.

II. Common Clinical Monitoring Parameters and Measurement Principles

Electrocardiogram

The principle of electrocardiogram monitoring is basically the same as the detection principle of conventional electrocardiographs. ECG monitors can generally monitor 3–6 leads, including standard leads Ⅰ, Ⅱ, Ⅲ, and pressurized leads aVR, aVL, and aVF, and can simultaneously display the waveforms of one or two leads. The powerful monitor can monitor 12 electrocardiogram leads. The simplest monitor typically has three monitoring electrodes.

The color identification of monitoring lead electrode includes American Heart Association (AHA) and International Electrotechnical Commission (IEC) standards, see Table 4-7.

Table 4-7　Color identification of monitoring lead electrodes

Standard	Electrode				
	Right arm −R, right upper chest	Left arm −L, left upper chest	Left leg −F, left lower chest	Right leg −N, right lower chest	Chest or V1−V6
AHA	White	Black	Red	Green	Brown
IEC	Red	Yellow	Green	Black	White

When the monitor has three monitoring electrodes, the monitoring electrodes are placed in the chest position, as shown in Fig. 4-6.

ECG monitors usually have the following selected functions: arrhythmia detection, S-T segment analysis, recall waveform display, run chart analysis, electrode shedding alarm, power failure processing, data storage and transmission, etc. Multiple channels can record multiple leads simultaneously.

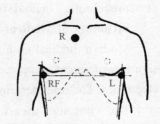

Fig. 4-6　Schematic diagram of electrocardiogram monitoring

Note: Although the detection principle of electrocardiogram monitoring is basically the same as that of conventional electrocardiographs, the electrocardiogram monitoring function cannot completely replace conventional electrocardiographs. At present, the electrocardiogram waveform of monitors generally cannot provide more detailed structures.

Heart Rate

Heart rate refers to the number of beats per minute of the heart. The average heart rate of healthy adults in a quiet state is 75 beats/min, with a normal range of 60-100 beats/min. Under different physiological conditions, the heart rate can reach a minimum of 40 beats/min and a maximum of 200 beats/min. The heart rate alarm range of the monitor: the low limit is 20-100 beats/min, and the high limit is 80-240 beats/min. Heart rate measurement is the measurement of instantaneous heart rate and average heart rate based on electrocardiogram waveforms.

Instantaneous heart rate refers to the reciprocal of two adjacent R-R intervals in the electrocardiogram. Namely:

$$F = 1/T \text{(times/s)} = 60/T \text{(times/min)}$$

where T is the R-R interval.

Invasive Blood Pressure

1. The Concept of Invasive Blood Pressure

Use catheterization to measure and monitor arterial blood pressure, central pulse pressure, left atrial pressure, left ventricular pressure, pulmonary artery and pulmonary capillary wedge pressure, etc. After percutaneous arterial puncture and indwelling of the catheter, it is fixed, and the pressure measuring tube is connected to the monitor through a transducer, which can display the arterial pressure waveform.

The most commonly used is the radial artery, followed by the femoral artery or brachial artery, which can not only continuously measure pressure, but also repeatedly collect blood samples to detect blood gas.

2. Characteristics and Significance of Various Blood Pressure Waveforms

Right atrial pressure (RAP) also represents central venous pressure (CVP), which is normally 6 – 12 cmH$_2$O (1 cmHg=13.332 2 Pa).

Right ventricular pressure (RVP), normal systolic blood pressure 20 – 30 mmHg, diastolic blood pressure 0 – 5 mmHg, and end-diastolic blood pressure 2 – 6 mm Hg.

Right ventricular pressure (RVP), normal systolic blood pressure 20 – 30 mmHg, diastolic blood pressure 0 – 5 mmHg, and end-diastolic blood pressure 2 – 6 mm Hg.

Pulmonary artery pressure (PAP), systolic blood pressure 20 – 30 mmHg, end-diastolic blood pressure 8 – 12 mm Hg, and average blood pressure 10 – 20 mmHg.

Pulmonary artery wedge pressure (PWP) is the pressure measured by the distal catheter after inflation of the balloon, which is the pressure formed in the reverse direction of the left atrium. When pulmonary resistance is normal, the secondary pressure is equal to the left atrium pressure.

3. Non-invasive Blood Pressure Measurement

In order to reduce the pain caused by the puncture band and reduce the possibility of blood pollution, non-invasive blood pressure measurement technology and methods have been introduced through scientific research and clinical application. The non-invasive blood pressure testing is shown in Fig. 4–7.

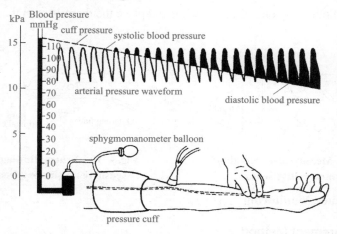

Fig. 4–7　Schematic diagram of non-invasive blood pressure testing

Blood Oxygen Saturation

The effective oxygen molecules in the blood form oxygenated hemoglobin (HbO_2) by binding to hemoglobin (Hb). The percentage of oxygenated hemoglobin in total hemoglobin is called blood oxygen saturation. Blood oxygen saturation is an important parameter that measures the ability of human blood to carry oxygen. Fig 4-8 shows the blood oxygen saturation test.

Fig. 4-8　Blood oxygen saturation test

Respiratory Rate

Respiratory monitoring refers to monitoring the patient's respiratory rate, also known as respiratory rate. Respiratory rate is the number of times a patient breathes per unit of time, measured in breaths/ min.

When breathing calmly, newborns should take 60-70 breaths per minute and adults should take 12-18 breaths per minute.

Thermal respiratory measurement:

Placing a thermistor in the nasal cavity changes the heat transfer conditions when the respiratory airflow passes through the thermistor, causing the temperature of the thermistor to change periodically with the respiratory airflow cycle, resulting in periodic changes in the thermistor value (Fig. 4-9).

R_1 and R_2 in Fig 4-9 are standard resistors, and $R_1=R_2$; R_4 is the zero adjustment resistance. Before use, adjust R_4 to make it equal to the thermistor value. The bridge is in balance and the output is zero. R_s is used to adjust the sensitivity of the bridge.

Fig. 4-10 is a schematic diagram of a thermosensitive respiratory frequency sensor. The thermistor is placed on the outside of the front end of the straight piece of the clamp. When using, simply clip the clip onto the nose wing and place the thermistor in the nostril.

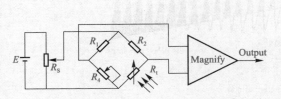

Fig. 4-9　Measurement circuit of thermosensitive sensor

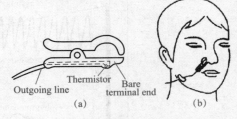

Fig. 4-10　Schematic diagram of thermosensitive respiratory frequency sensor

Temperature Monitoring

1. Measurement Method

The temperature measurement in the monitor generally uses the thermistor with negative temperature coefficient as the temperature sensor. The measurement circuit for body temperature measurement is a Wheatstone bridge, where a thermistor is connected to one arm of the bridge. By measuring the unbalanced output of the bridge, body temperature can be measured.

2. Factors Affecting Body Temperature Measurement

A thermometer should be able to provide fast, accurate, and reliable body temperature measurement. The factors that affect body temperature measurement include the following items.

(1) Scale accuracy;
(2) There is no appropriate reference standard to calibrate the thermometer;
(3) Selection of anatomical sites for measurement;
(4) Environmental factors;
(5) Patient activity and movement.

End-Tidal Carbon Dioxide Monitoring

End-tidal carbon dioxide ($ETCO_2$) is an important detection indicator for anesthesia patients and patients with respiratory and metabolic diseases. Detecting the concentration of end-respiratory carbon dioxide not only detects ventilation but also reflects pulmonary blood flow, which has the advantages of non-invasive and continuous monitoring, thereby reducing the frequency of blood gas analysis.

There are two types of CO_2 monitoring: mainstream and bypass.

(1) The mainstream method directly places the gas sensor in the patient's respiratory duct, directly converts the concentration of CO_2 in the respiratory gas, and then sends the electrical signal to the monitor for analysis and processing to obtain $ETCO_2$ parameters.

(2) A side flow optical sensor is placed inside the monitor, and a gas sampling tube is used to extract a patient's respiratory gas sample in real time. When removing the breath, a patient's respiratory gas sample is taken, and the moisture in the respiratory gas is removed through the oxygen water separator. The sample is then sent to the monitor for CO_2 analysis.

Pulse

Pulse refers to the periodic pulsation of arterial blood vessels as the heart relaxes and contracts. Pulse includes changes in various physical quantities such as intravascular pressure, volume, displacement, and wall tension.

There are several methods for measuring pulse:

(1) Extracting from electrocardiogram signals;
(2) Calculate pulse rate based on the fluctuations detected by the pressure sensor during blood pressure measurement;
(3) Photocapacitance method (Fig. 4-11).

Pulse is a signal that changes periodically with the beating of the heart, and the volume

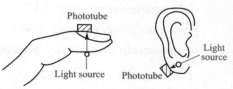

Fig. 4-11 Pulse measurement using photocapacitance volumetric method

of arterial blood vessels also changes periodically. The electrical cycle of the photoelectric converter is the pulse rate.

【Words and phrases】

monitor	n. 监护仪；显示器；班长；
	v. 监视；监听
physiological parameter	生理参数
display	n. 显示；陈列；
	n. 表演；展览；显示
electrocardiograph (ECG)	心电图
respiration	n. 呼吸；（医）一次呼吸
pulse rate	脉搏率：心脏每分钟跳动的次数
non-invasive blood pressure (NIBP)	无创血压（NIBP）
carbon dioxide (CO_2)	二氧化碳
anesthetic gas (AG)	麻醉气体
impedance cardiograph (ICG)	阻抗心电图
bispectral index (BIS)	双光谱指数
respiration mechanics (RM)	呼吸力学（原理）
pediatric	adj. 儿科的
neonatal	adj. 新生的；初生的
physiological alarm area	生理报警区
patient demographics	病人的人口统计
kidney dialysis	肾脏透析
hyperbaric oxygen chamber	[临床] 高压氧舱
radiation therapy	放射治疗：一种使用高能辐射来杀死或控制癌细胞的治疗方法。
mental illness	精神疾病：一类医学疾病，主要表现为个性、心智或情绪的严重紊乱
preterm infant monitor	早产儿监护仪
blood oxygen saturation	血氧饱和度：血液中氧气与血红蛋白结合的程度
American Heart Association (AHA)	美国心脏协会
International Electrotechnical Commission (IEC)	国际电工委员会
right ventricular pressure (RVP)	右心室压（RVP）
pulmonary artery pressure (PAP)	肺动脉压（PAP）
thermosensitive respiratory frequency sensor	热敏电阻呼吸频率传感器

 Course practice

1. Complex Sentence Analysis.

(1) This patient monitor is intended to be used for monitoring, displaying, reviewing, storing and transferring of multiple physiological parameters including electrocardiogram (ECG), heart rate (HR), respiration (Resp), temperature (Temp), SpO_2....

① be intended to + do sth：用来做某事，为了；+ to be：规定为，确定为。

② be used for + doing：被用来做某事，用来做某事；+for used as 被用来做。

【译文】医用监护仪用于监测、显示、查看、存储和传输多种心理参数，包括心电图（ELG）、心率（HR）、呼吸（Resp）、体温（Temp）、SpO_2……

(2) Interpretation of resting 12-lead ECG and CO monitoring are restricted to adult patients only.

be restricted + to sth.：限于，被限制在……范围内；+ by sth.：被……限制。

【译文】静息12导联心电图和心输出量监测的解释仅限于成人患者。

(3) The arrhythmia detection, ECG ST-segment, BIS and RM monitoring are not intended for neonatal patients.

be intended for：打算为……所用，预定给。

【译文】心律失常监测、ECG ST段、BIS和RM监测不适用于新生儿患者。

(4) The safety performance is in line with international standards.

be in line with：符合，与……一致，标识某事物与另一事务在目标、原则上一致。

"符合"的其他英语表达：accord with（与某物相符）；conform to（与某规范相符）。

【译文】安全性能符合国际标准。

(5) Instantaneous heart rate refers to the reciprocal of two adjacent R-R intervals in the electrocardiogram.

① refer to：指的是，关联，引用；+ as：称作。

② the reciprocal of：…的倒数；reciprocal：*adj.* 相互的，互惠的；反向的，倒数的。

③ instantaneous：*adj.* 瞬间发生的，瞬间完成的。

【译文】瞬时心率是指心电图中两个相邻的R-R间隔的倒数。

2. Translate the following sentences.

(1) The use of medical devices is for diagnosis, surgery, treatment, laboratory testing, monitoring, experimental equipment, human organ function replacement, and so on.

(2) Most of these functions are targeted at medical institutions and medical teaching and research institutions, so their users are also the main body of these institutions.

3. Choose the proper answer to fill in the blank and translate the sentences.

(to, in, of, at, from, on)

(1) Due ( ) the high technological content in the manufacturing of medical instruments and equipment...

(2) ( ) particular, the understanding and accuracy of the instructions, labels, and packaging labels of medical device products are not accurate.

(3) The main problems are that there is no record ( ) purchasing medical devices and the legal certification and qualification of the supplier cannot be provided.

(4) Interpretation of resting 12-lead ECG and CO monitoring are restricted ( ) adult patients only.

Translation of the text

监护仪

I. 医用监护仪概述

监护仪的概念

医用监护仪能够对人体的生理参数进行长时间连续监测，并且能对监测结果进行存储、显示、分析和控制，出现异常情况时便发出警报提醒医护人员及时进行处理（图4-4）。

图4-4 监护仪示意

监护仪的用途

医用监护仪用于监测、显示、查看、存储和传输多种生理参数，包括心电图（ECG）、心率（HR）、呼吸（Resp）、体温（Temp）、SpO_2、脉搏率（PR）、无创血压（NIBP）、有创血压（IBP）、心输出量（CO）、二氧化碳（CO_2）、氧气（O_2）、麻醉气体（AG）、心阻抗图（ICG），成人、儿童和新生儿患者的双频谱指数（BIS）和呼吸力学（RM）。

静息12导联心电图和心输出量监测的解释仅限于成年患者。ICG仅适用于符合以下要求的成人患者：身高为122～229 cm，体重为30～159 kg。心律失常监测、ECG ST段、BIS和RM监测不适用于新生儿患者。

该监护仪将由临床专业人员或在其指导下在医疗机构中使用。它不适用于直升机运输、医院救护车或家庭使用。

监护仪的显示屏介绍

监护仪采用高分辨率 TFT LCD 显示患者参数和波形。图 4-5 所示为一个典型的监护仪显示界面。

1. 患者信息区

此区域显示患者信息，如科室、床位号、患者姓名、患者类别和治疗进度。

：表示没有患者入院或患者信息不完整。

：表示患者有起搏器。

如果没有患者入院，选择此区域将进入"患者设置"菜单。如果患者已经入院，选择此区域将进入"患者人口统计"菜单。

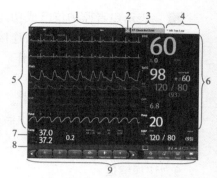

图 4-5 典型的监护仪显示界面
1—患者信息区；2—报警符号区；
3—技术报警区；4—生理报警区；
5—波形区；6—参数区 A；
7—参数区 B；8—提示消息区；
9—快捷键区

2. 报警符号区

：表示报警已暂停。

：表示报警声音暂停。

：表示报警声音已关闭。

：表示系统处于报警关闭状态。

3. 技术报警区

此区域显示技术报警消息和提示消息。当收到多条消息时，它们将循环显示。选择此区域，将显示技术报警列表。

4. 生理报警区

此区域显示生理报警信息。当发生多个报警时，它们将循环显示。选择此区域，将显示生理报警列表。

5. 波形区

此区域显示测量波形，波形名称显示在波形的左上角。选择此区域，将显示相应的测量设置菜单。

6. 参数区 A

此区域显示测量参数。每个监测的参数都有一个参数窗口，参数名称显示在左上角。每个参数的对应波形显示在波形区域的同一行中。选择此区域，将显示相应的测量设置菜单。

7. 参数区 B

对于该区域中显示的参数，不显示其对应的波形。

8. 提示消息区

此区域显示提示消息、网络状态图标、电池状态图标、日期和时间等。

▇：表示患者监护仪已成功连接到有线网络。
▇：表示患者监护仪无法连接有线网络。
▇：表示无线功能正在工作。
▇：表示无线功能不工作。
▇：表示插入了 CF 存储卡。
▇：表示连接了辅助显示器或远程显示器。
▇："屏幕设置"按钮。

9．快捷键区

该区域包含快捷键，可让您快速访问各项功能。

医用监护仪的临床应用

根据临床护理对象和监护目的不同，临床监护仪用于以下护理监护：手术中和手术后的监护；分娩监护和胎儿监护；危重患者的监护；恢复期患者的监护；治疗患者（肾透析、高压氧舱、放射线治疗、精神病等）的监护。

根据需要在科室和病房中使用各种专用的监护仪（也称为监护系统），主要包括危重患者监护仪、冠心病监护仪、分娩监护仪、新生儿和早产儿监护仪、颅内压监护仪、麻醉监护仪、睡眠监护仪等。

监护仪监测的生理和生化参数有 ECG 监测、呼吸监测、无创血压监测、有创血压监测、气道二氧化碳监测、气道氧气监测、血容量辅助监测、血氧饱和度/血液体积监测、温度监测、pH 值监测、血气监测等。医用监护仪除有监护功能外，还有疾病诊断和治疗的功能，同时还有抢救功能，如动态心电图（Holter）和血压监护仪、心脏除颤监护仪等。

医用监护仪的结构

现代医用监护仪器主要由四个部分组成，即生命体征测量组件、主机及系统、连接接口和传感器与电极。各部分的功能简要描述如下。

生命体征测量组件主要包含如心电、呼吸、体温、血压、脉搏氧饱和度、呼吸末二氧化碳等人体生命体征的各种测量组件。

主机及系统主要包含主控板、显示器、键盘、记录仪及运行软件、其他外部扩展设备。

生命体征测量组件与主机接口有双向通信的需求，因此需要对每个生命体征测量模块设置特征识别，便于主机进行有针对性的操作。

传感器、连接电缆与生命体征测量组件的连接形式分为硬件连接和智能连接。

医用监护仪的特点

（1）安全性能符合国际标准。
（2）功能更加强大，性能更加卓越。
（3）专用监护仪发展迅速。
（4）远程监护和家庭监护日益普及。

Ⅱ. 临床常用的监护参数及测量原理

心电图

心电监护原理与常规心电图机的监测原理基本相同。ECG 监护仪一般能监护 3～6 个导联，标准 Ⅰ、Ⅱ、Ⅲ 导联及加压导联 aVR、aVL、aVF，能同时显示其中的一个或两个导联的波形。功能强大的监护仪可监护 12 个心电导联。最简单的监护仪一般有 3 个监护电极。

监护导联电极的颜色标识有美国心脏协会（American Heart Association，AHA）和国际电工委员会（International Electrotechnical Commission，IEC）两个标准，如表 4-7 所示。

表 4-7 监护导联线电极的颜色标识

标准	电极				
	右臂 R、右上胸部	左臂 L、左上胸部	左腿 F、左下胸部	右腿 N、右下胸部	胸部或 V1~V6
AHA	白色	黑色	红色	绿色	棕色
IEC	红色	黄色	绿色	黑色	白色

当监护仪有 3 个监护电极时，监护电极放置于胸部的位置，如图 4-6 所示。

ECG 监护仪通常还有以下备选功能：心律不齐检测、S-T 段分析、回忆波形显示、趋势图分析、电极脱落报警、电源故障处理、数据存储和传送等。可以有多个通道同时记录多个导联。

注意：虽然心电监护原理与常规心电图机的监测原理基本相同，但心电监护功能并不能完全替代常规心电图机。目前监护仪的心电波形一般不能提供更详细的结构。

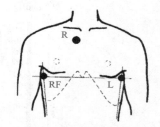

图 4-6 心电监护示意

心率

心率是指心脏每分钟搏动的次数。健康的成年人在安静状态下平均心率为 75 次/min，正常范围为 60～100 次/min。在不同生理条件下，心率最低可为 40 次/min，最高可达 200 次/min。监护仪心率报警范围：低限为 20～100 次/min，高限为 80～240 次/min。心率测量是根据心电波形测定瞬时心率和平均心率。

瞬时心率是指心电图中两个相邻的 R-R 间隔的倒数。即

$$F = 1/T （次/s） = 60/T （次/min）$$

式中，T 为 R-R 间期。

有创血压

1. 有创血压的概念

有创血压指利用导管术来测量和监护动脉血压、中心脉冲压、左心房压、左心室压、肺动脉和肺毛细血管楔压等。经皮动脉穿刺并留置导管后，导管被固定，测

压管通过换能器连接于监视器，可显示动脉压力波形。

最常用的是桡动脉，其次是股动脉或肱动脉，不仅可以连续测压，还可反复采集血标本以检测血气。

2．各血压波形的特点及意义

右房压（RAP），也代表中心静脉压（CVP），正常为 6～12 cmH$_2$O（1 cmHg=13.332 2 Pa）。

右室压（RVP），正常收缩压为 20～30 mmHg，舒张压为 0～5 mmHg，舒张末压为 2～6 mmHg。

右室压（RVP），正常收缩压为 20～30 mmHg，舒张压为 0～5 mmHg，舒张末压为 2～6 mmHg。

肺动脉压（PAP），收缩压为 20～30 mmHg，舒张末压为 8～12 mmHg，平均压为 10～20 mmHg。

肺动脉楔压（PWP），即通过气囊充气后由远端导管测得的压力，测得的是左心房逆向形成的压力，在肺阻力正常时次压力与心左房压力相等。

3．无创血压测量

为降低穿刺带给病人的痛苦及降低血液污染的可能性，经过科研与临床应用，又推出了无创血压测量的技术与方法。无创血压测试如图 4-7 所示。

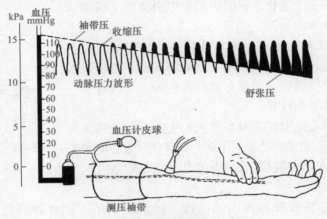

图 4-7 无创血压测试示意

血氧饱和度

血液中的有效氧分子，通过与血红蛋白（Hb）结合后形成氧合血红蛋白（HbO$_2$），氧合血红蛋白占全部血红蛋白的百分比称为血氧饱和度。血氧饱和度是衡量人体血液携氧能力的重要参数。图 4-8 所示为血氧饱和度测试示意。

呼吸率

呼吸监护指监护患者的呼吸频率，即呼吸率。呼吸率是患者在单位时间内呼吸的次数，单位是次每分（次/min）。

图 4-8 血氧饱和度测试示意

平静时，新生儿呼吸率为 60～70 次/min，成人为 12～18 次/min。

热敏式呼吸率测量：将热敏电阻置于鼻腔内，当呼吸气流通过热敏电阻时，改变了传热条件，使热敏电阻的温度随呼吸气流发生变化，从而使热敏电阻值发生周期性的变化（图 4-9）。

在图 4-9 中，R_1、R_2 为标准电阻，且 $R_1=R_2$；R_4 为调零电阻，使用前调节 R_4 使它与热敏电阻值相等，电桥处于平衡状态，输出为零。R_S 是用来调整电桥灵敏度的。

图 4-10 是热敏电阻式呼吸频率传感器示意。热敏电阻放在夹子的平直片前端外侧。使用时只要用夹子夹住鼻子翼，并使热敏电阻置于鼻孔之中即可。

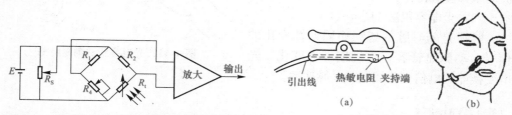

图 4-9　热敏电阻式呼吸频率传感器的测量电路　　图 4-10　热敏电阻式呼吸频率传感器示意

体温监测

1. 测量方法

监护仪中的体温测量一般采用负温度系数的热敏电阻作为温度传感器。体温测量的测量线路是惠斯通电桥，将热敏电阻接在电桥的一个桥臂上，通过测量电桥的不平衡输出，可以监测体温。

2. 影响体温测量的因素

体温计应该能够提供快速、准确、可靠的体温测量。影响体温测量的因素包括以下几个。

（1）刻度准确性；

（2）是否有适当的参考标准来对体温计进行校准；

（3）测量的解剖部位的选择；

（4）环境因素；

（5）患者的活动和移动。

呼吸末二氧化碳监测

呼吸末二氧化碳（ETCO$_2$）是麻醉患者和呼吸代谢系统疾病患者的重要检测指标。检测呼吸末二氧化碳浓度，不仅可检测通气而且能反映肺血流量，具有无创、连续监测的优点，从而减少血气分析的次数。

CO_2 监护有主流式和旁流式两种。

（1）主流式直接将气体传感器放置在患者呼吸气路导管中，直接对呼吸气体中的 CO_2 进行浓度转换然后将电信号送入监护仪进行分析处理，得到 ETCO$_2$ 参数。

（2）旁流式的光学传感器置于监护仪内，由气体采样管实时抽取患者呼吸气体样品。呼吸时，取患者呼吸气体样品，经氧水分离器，去除呼吸气体中的水分，送入监护仪中进行 CO_2 分析。

脉搏

脉搏是动脉血管随心脏舒张收缩而周期性搏动的现象。脉搏包含血管内压、容积、位移和管壁张力等多种物理量的变化。

脉搏测量的方法有：

（1）从心电信号中提取；

（2）从测量血压时压力传感器测到的波动来计算脉搏率；

（3）光电容积法（图4-11）。

脉搏是随心脏的搏动而周期性变化的信号，动脉血管容积也周期性地变化，光电变换器的电信变化周期就是脉搏率。

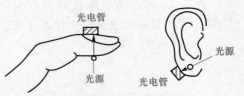

图4-11 光电容积法测量脉搏

 Evaluate

任务名称		医用监护仪	姓名	组别	班级	学号	日期	
考核内容及评分标准			分值	自评	组评	师评	平均分	
三维目标	知识	了解医用监护仪的概念及作用等	25分					
	技能	能阅读专业英语文章、翻译专用英语词汇	40分					
	素养	在学习过程中秉承科学精神、合作精神	35分					
加分项	收获（10分）	收获（借鉴、教训、改进等）：	你进步了吗？		加分			
			你帮助他人进步了吗？					
	问题（10分）	发现问题、分析问题、解决方法、创新之处等：			加分			
总结与反思					总分			

2.2 Defibrillator

Text

Concept of Defibrillator

In a very short period of time, a strong current is supplied to the heart (currently, direct current is used), which can depolarize all cardiac autonomic cells at the same time in an instant, and inactivate all possible reentry channels. Then the sinoatrial node, which has the highest autonomy in the cardiac pacing system, can regain its dominant position to control the heart beat, so the heart rhythm returns to sinus.Fig. 4-12 and Fig. 4-13 are two physical images of defibrillators.

Fig. 4-12　Physical image of defibrillator 1

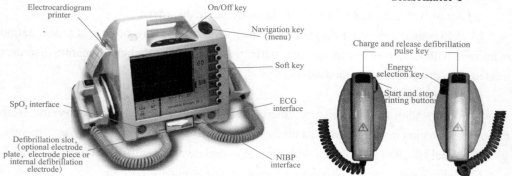

Fig. 4-13　Physical image of defibrillator 2

Types of Defibrillators

1. Unidirectional Waveform Defibrillator [Fig. 4-14 (a)]

The energy selection is from 0 to 360 J, and different energies are selected based on the patient's age and weight adjustment.

2. Bidirectional Waveform Defibrillator [Fig. 4-14 (b)]

The energy selection is from 0 to 200 J, which is safer and more effective than unidirectional waveform defibrillators. Therefore, bidirectional waveform defibrillators are now commonly used in clinical practice.

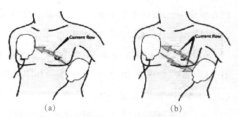

Fig. 4-14　Two methods of defibrillation
(a) Monophasic; (b) Biphasic

Types of Defibrillation

1. Synchronous Defibrillation

It is used for cardioversion of atrial fibrillation, flutter, supraventricular and ventricular tachycardia. Using the R wave on the patient's electrocardiogram to trigger discharge, the electrical pulse occurs at the descending branch of the R wave. Energy selection starts from low. Monitor the patient's ECG during defibrillation. If it does not revert to sinus rhythm, electric power can be increased and do electric defibrillation again.

2. Non-synchronous Defibrillation

It is only used for ventricular fibrillation and flutter. At this time, the patient has lost consciousness, and should be immediately defibrillated. After defibrillation, observe whether the patient's heart rhythm changes to sinus rhythm through the electrocardiograph. Multiple intermittent defibrillation is possible, with energy ranging from low to high.

(1) Indications.

① Ventricular fibrillation and flutter are absolute evidences of electric defibrillation.

② Atrial fibrillation and flutter with hemodynamic disorder.

③ Paroxysmal supraventricular tachycardia, ventricular tachycardia and preexcitation syndrome with tachyarrhythmia that are ineffective or have severe hemodynamic disorders treated with drugs and other methods.

(2) Contraindications.

① With a long medical history, there is significant cardiac hypertrophy and fresh thrombus formation in the atrium, or a history of thrombus formation within the past 3 months.

② Atrial fibrillation or flutter with high or complete atrioventricular block.

③ Heterotopic tachyarrhythmia with sick sinoatrial node syndrome.

④ When there is a history of digitalis poisoning and hypokalemia, it is not suitable for electrical defibrillation.

Use of Defibrillators

1. Non-synchronous Electric Defibrillation

If the patient experiences ventricular fibrillation or flutter, electric defibrillation should be performed as early as possible. The early onset of ventricular fibrillation is generally coarse fibrillation, and defibrillation is easy to succeed at this time, so it should be attempted within 2 min.

Defibrillators are commonly used in emergency situations for patients with cardiac arrest, as shown in Fig. 4-15. The curve between defibrillation time and patient survival rate is shown in Fig. 4-16.

Fig. 4-15 Rescue the patient

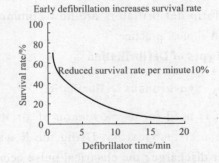

Fig. 4-16 Defibrillation survival curve

Operating steps are as below.

(1) Prepare for defibrillation, and prepare various rescue equipment and drugs:

defibrillator, electrocardiograph, oscilloscope, and rescue equipment and drugs required for cardiopulmonary resuscitation.

(2) The patient is lying flat on a hard bed, with an open venous passage that fully exposes the chest wall.

(3) Perform routine electrocardiogram and electrocardiogram monitoring before surgery. After completing the electrocardiogram recording, remove the lead wire from the electrocardiograph to avoid electric shock damage to the electrocardiograph.

(4) Select a suitable electrode plate and evenly apply conductive adhesive. Clean the skin of the patient's defibrillation site with a piece of physiological saline, water, and gauze. Apply the negative electrode (apex) to the left midline and three transverse fingers under the armpit (apex of the heart). At the positive electrode (stenal): at the right edge of the right subclavian sternum (at the base of the heart).

(5) Select asynchronous electric defibrillation.

(6) Choose the appropriate energy: the unidirectional waveform defibrillator is generally chosen for adults at 300-360 J, and for children at 2 J/kg of body weight. Adults generally choose a bidirectional waveform defibrillator of 150-200 J, while children weigh 2 J/kg.

(7) Charging: Before charging, instruct other personnel not to come into contact with the patient, the hospital bed, and the instruments and equipment connected to the patient to avoid electric shock. Then, press the charging switch, and the screen will display the predetermined energy to be fully charged.

(8) After charging, place the two electrodes correctly on the patient's skin according to the diagram on the electrode, and apply appropriate pressure to ensure that the electrode plate is in good contact with skin. Press the discharge button on the electrode plate with both hands and thumbs at the same time (Fig.4-17).

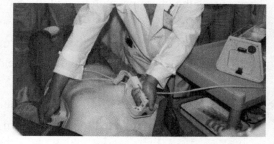

Fig. 4-17 Actual application scenarios of defibrillation

(9) After defibrillation, immediately observe whether the patient's ECG reverts to sinus rhythm.

(10) If ventricular fibrillation, flutter, etc., continue to occur and defibrillation fails, it should be recharged and the steps repeated after a certain interval of time.

(11) After the operation is completed, reset the energy switch to zero, position the patient, monitor heart rate and rhythm, and follow the doctor's instructions for medication.

(12) Keep records.

2. Synchronous Electric Defibrillation

If the patient has atrial fibrillation, flutter, supraventricular and ventricular tachycardia, synchronous defibrillation is feasible.

Operating steps are as below.

(1) Those who chose to have defibrillation stopped using digitalis for 1-3 days before operation and were given drugs to improve heart function and correct hypokalemia and acidosis.

(2) Take quinidine, procainamide and other drugs 1-2 days before defibrillation to prevent arrhythmia recurrence after defibrillation.

(3) On the day of defibrillation, fast and empty the bladder.

(4) Establish a venous channel, connect an electrocardiogram machine and an electrocardiogram monitor, perform a full lead electrocardiogram before surgery, and select a lead with a larger R wave to test the synchronization of electrical defibrillation.

(5) Propofol, etomidate and other anesthetic should be properly used, or diazepam should be slowly injected intravenously until the patient's eyelash reflex begins to disappear during the deep substitute anesthesia process, and breathing should be closely observed. If there is respiratory depression, oxygen should be given to the mask.

(6) Select synchronous electric defibrillation.

(7) Select appropriate energy, atrial fibrillation is 150 - 200 J, atrial flutter is 80 - 100 J, and supraventricular tachycardia is 100 J.

(8) Charging: Before charging, instruct other personnel not to come into contact with the patient, hospital bed, or instruments connected to the patient to avoid electric shock. Then, press the charging switch, and the screen will display the predetermined energy to indicate full charge.

(9) After charging, place the two electrodes correctly on the patient's skin according to the diagram on the electrode, and press the discharge button on the electrode board with both hands and thumbs simultaneously.

(10) After discharge, observe whether the patient's ECG has reverted to sinus rhythm. If it does not reverted to sinus rhythm, increase the electric power and restore the rhythm again.

(11) After the operation is completed, return the energy switch to the zero position, measure the heart rhythm, and follow the doctor's instructions for medication.

(12) Keep records.

Precautions

(1) The defibrillator should be regularly checked for performance, charged in a timely manner, and disinfected after use.

(2) When applying conductive paste to the electrode plate, it is not allowed to rub the two electrode plates against each other. The electrode plate should be in close contact with the patient's skin to ensure good conductivity.

(3) If the patient is thin or has uneven skin, two gauze blocks dipped in physiological saline can be placed directly on the patient's defibrillation site.

（4）During an electric shock, no one should come into contact with the patient or hospital bed to avoid electric shock.

（5）For patients with fine fibrillation type ventricular fibrillation, cardiac compression, oxygen therapy, and medication should be performed first to turn it into coarse fibrillation, followed by electric shock to improve the success rate.

（6）The skin at the shock site may have mild erythema, pain, or myalgia, which can be relieved by itself after 3–5 days.

（7）During open chest defibrillation, the electrodes are placed directly on the front and rear walls of the heart, and the defibrillation energy is generally 5–10 J.

（8）For ventricular tachycardia that can clearly distinguish QRS and T waves, synchronous electrical defibrillation should be performed. For those that cannot be distinguished, asynchronous electrical defibrillation should be used.

【Words and phrases】

defibrillator	n. 除颤器（通过电击心脏控制心肌运动）
depolarize	v. 去极化；去（退）极（化）；俏偏振（光）；去（退）磁；扰乱；动摇；消解；使丧失信心
rhythm	n. 节奏；节律；韵律；规律；律动；规则变化
bidirectional	adj. 双向的
clinical	adj. 临床的；临床诊断的
synchronous	adj. 同步的；共时的；同时发生（或存在）的
revert	v. 恢复；回复（到以前的状态、制度或行为）；还原；回到
conductive	adj. 导电（或热等）的；能传导（电、热等）的
substitute	v.（以……）代替；代用；接替；替代；＜化＞（使）取代；派……接替，由……替代
	n. ＜语＞替代（形式）；代用词；代用语；代用品；代替者；替补队员；代替者；替换者

Course practice

1. Complex sentence analysis.

（1）If it does not revert to sinus rhythm, electric power can be increased and do electric defibrillation again.

do not revert to：不要返回到 / 恢复到。

【译文】若不能恢复窦性心律，可增加电功率，再次除颤。

（2）When there is a history of digitalis poisoning and hypokalemia, it is not suitable for electrical cardioversion.

is not suitable for...：不适合……

【译文】当有洋地黄中毒史和低钾血症史时，暂不宜用电复律。

（3）The early onset of ventricular fibrillation is generally coarse fibrillation, and defibrillation is easy to succeed at this time, so it should be attempted within 2 min.

① be attempted：试图/尝试。

② coarse fibrillation：粗颤。

【译文】室颤发生的早期一般为粗颤，此时除颤易成功，故应争取在2 min内进行。

（4）Charging: Before charging, instruct other personnel not to come into contact with the patient...

come into contact with...：接触……。指与某物或某人直接或间接地相互作用或沟通。

【译文】充电：充电前，提醒其他人员不要与患者接触。

2. Translate the following paragraphs.

（1）It is only used for ventricular fibrillation and flutter. At this time, the patient has lost consciousness, and should be immediately defibrillated. After defibrillation, observe whether the patient's heart rhythm changes to sinus rhythm through the electrocardiograph.

（2）Multiple intermittent defibrillation is possible, with energy ranging from low to high.

3. Choose the proper answer to fill in the blank and translate the sentences.

(as, from, to, at, of, in)

（1）If the patient experiences ventricular fibrillation or flutter, electric defibrillation should be performed as early（　　）possible.

（2）After completing the electrocardiogram recording, remove the lead wire（　　）the electrocardiograph.

（3）After the operation is completed, reset the energy switch（　　）zero, position the patient.

 Translation of the text

除颤仪

除颤仪的概念

在极短暂的时间内给心脏通以强电流（目前用直流电），可使所有心肌在瞬间同时除极，并使所有可能存在的折返通道全部失活。然后心脏起搏系统中具有最高自律性的窦房结可以恢复主导地位控制心搏，于是心律转复为窦性。图4-12和图4-13为两款除颤仪实物图。

图4-12　除颤仪实物图1

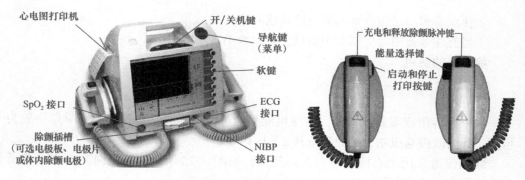

图 4-13 除颤仪实物图 2

除颤仪的种类

1. 单向波形除颤仪[图 4-14（a）]

能量选择为 0～360 J，根据患者年龄、体重，选择不同的能量。

2. 双向波形除颤仪[图 4-14（b）]

能量选择为 0～200 J，比单向波形除颤仪更加安全有效，故现在临床多已使用双向波形除颤仪。

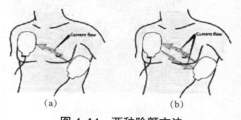

图 4-14 两种除颤方法
(a) 单相的；(b) 双相的

除颤的种类

1. 同步电除颤

同步电除颤用于心房颤动、扑动，室上性及室性心动过速等的复律。利用患者心电图上的 R 波触发放电，其电脉冲发生在 R 波降支。能量选择从低开始。除颤时监测患者 ECG。若未转复为窦性心律，可增加电功率，再次电除颤。

2. 非同步电除颤

非同步电除颤仅用于心室颤动和扑动。在患者神志多已丧失时，应立即除颤，除颤后通过心电示波器观察患者心律是否转为窦性。可多次间断除颤，能量由低到高。

（1）适应证。
①心室颤动、扑动是电除颤绝对指证。
②心房颤动和扑动伴血流动力障碍者。
③药物及其他方法治疗无效或有严重血流动力学障碍的阵发性室上心动过速、室性心动过速、预激综合征伴快速心律失常者。

（2）禁忌证。
①病史较长，心脏明显肥大及心房有新鲜血栓形成或近 3 个月内有血栓史。
②伴高度或完全房室传导阻滞的心房颤动或扑动。

③伴病态窦房结综合征的异位性快速心律失常。
④当有洋地黄中毒史和低钾血症史时，暂不宜用电除颤。

除颤仪的使用

1．非同步电除颤

如果患者出现室颤、室扑，应尽可能早地进行电除颤。室颤发生的早期一般为粗颤，此时除颤易成功，故应争取在 2 min 内进行。

除颤仪通常用于心搏骤停病人的急救，如图 4-15 所示。除颤时间与病人存活率曲线如图 4-16 所示。

图 4-15 抢救患者

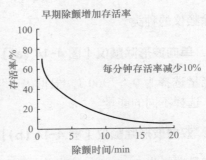

图 4-16 除颤存活率曲线

操作步骤如下。

（1）做好除颤准备，备好各种抢救器械和药品：除颤仪、心电图机、示波器、心肺复苏所需的抢救设备和药品。

（2）患者平卧于硬板床上，开放静脉通道，充分暴露胸部。

（3）术前进行常规心电图检查并进行心电监护。完成心电图记录后把导联线从心电图机上解除，以免电击损坏心电图机。

（4）选择合适电极板均匀涂抹导电胶，用生理盐水、清水、纱布清洁患者除颤部位的皮肤。贴负极（apex）处：左腋中线腋下三横指（心尖部）。贴正极（stenal）处：右锁骨下胸骨右缘（心底部）。

（5）选择非同步电除颤。

（6）选择合适的能量：单向波形除颤仪成人一般选择 300～360 J，儿童每千克体重选择 2 J。双向波形除颤仪成人一般选择 150～200 J，儿童每千克体重选择 2 J。

（7）充电：充电前叮嘱其他人员不得接触患者、病床以及与患者相连接的仪器设备以免触电，然后按下充电开关，屏幕显示到预定能量即为充满。

（8）充电完毕后，将两个电极按电极上图示正确地放在患者皮肤处，并施以适当压力使电极板与皮肤接触完好。双手大拇指同时按下电极板上的放电键（图 4-17）。

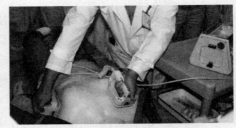

图 4-17 除颤仪实际应用场景

（9）除颤完毕，立即观察患者 ECG 是否转复为窦性心律。

（10）如果室颤、室扑等持续出现，复律失败，应重新充电，间隔一定时间后重复以上步骤。

（11）操作完成后，将能量开关回复至零位，安置好患者，监测其心律，并让患者遵医嘱用药。

（12）做好记录。

2．同步电除颤

若患者出现心房颤动、扑动、室上性及室性心动过速等，可进行同步电除颤。操作步骤如下。

（1）择期除颤的患者术前应停用洋地黄类药物 1～3 天，给予改善心功能、纠正低钾血症和酸中毒的药物。

（2）除颤前 1～2 天服用奎尼丁、普鲁卡因胺等药物，以防除颤后心律失常复发。

（3）除颤当天禁食，并排空膀胱。

（4）建立静脉通道，连接心电图机及心电监护仪，术前做全导心电图，选 R 波较大的导联测试电除颤的同步性。

（5）适当应用异丙酚、依托咪酯等麻醉药，或用地西泮缓慢静注，至患者睫毛反射开始消失，麻醉过程中严密观察患者呼吸，有呼吸抑制时，应给其戴上面罩给氧。

（6）选择同步电除颤。

（7）选择合适的能量，心房颤动为 150～200 J，心房扑动为 80～100 J，室上性心动过速为 100 J。

（8）充电：充电前，提醒其他人员不要与患者接触、病床以及与患者相连接的仪器设备以免触电，然后按下充电开关，屏幕显示到预定能量即为充满。

（9）充电完毕后将两个电极按电极上图示正确放在患者皮肤处，双手大拇指同时按下电极板上的放电键。

（10）放电完毕，观察患者 ECG 是否转复为窦性心律，若未转复为窦性心律，可增加电功率，再次复律。

（11）操作完成后，将能量开关回复至零位，监测患者心律，并让患者遵医嘱用药。

（12）做好记录。

注意事项

（1）除颤仪应定时检查除颤器性能，及时充电，使用后做好消毒处理。

（2）电极板涂导电胶时不可将两块电极板相互摩擦涂抹，电极板应与患者皮肤密切接触，保证导电良好。

（3）患者较瘦或皮肤不平整时，可将两块蘸有生理盐水的纱布直接放在患者除颤部位。

（4）电击时，任何人不得接触患者及病床，以免触电。

（5）对于细颤型室颤者，应先进行心脏按压，氧疗及药物治疗后，使之变为粗颤，再进行电击，以提高成功率。

（6）电击部位皮肤可能会有轻度红斑、疼痛，也可出现肌肉痛，3～5天后可自行缓解。

（7）开胸除颤时，电极直接放在心脏前后壁，除颤能量一般为5～10 J。

（8）对于能明确区分QRS波和T波的室速，应进行同步电除颤，无法区分者，采用非同步电除颤。

Evaluate

任务名称		除颤仪	姓名	组别	班级	学号	日期	
考核内容及评分标准				分值	自评	组评	师评	平均分
三维目标	知识	了解除颤的概念，除颤仪的定义、结构及分类等相关知识		25				
	技能	能阅读专业英语文章、翻译专用英语词汇		40				
	素养	在学习过程中秉承科学精神、合作精神		35				
加分项	收获（10分）	收获（借鉴、教训、改进等）：	你进步了吗？				加分	
			你帮助他人进步了吗？					
	问题（10分）	发现问题、分析问题、解决方法、创新之处等：					加分	
总结与反思							总分	

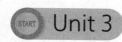 **Unit 3 Other Equipment**

Unit objectives

【Knowledge goals】
1. 掌握可变径钻孔、造穴、卸压、增透一体化装备的组成及功能等相关知识。
2. 掌握车载散热器类型、组成和工作原理等相关知识。

【Skill goals】
1. 能够阅读专业英语文章。
2. 可以翻译专用英语词汇。

【Quality goals】
1. 在学习过程中发扬科学研究精神。
2. 增强团队合作意识。

3.1 Tunnel Construction Equipment

 Text

其他工程机械

Variable Diameter Integration Equipment for Drilling, Cavitation, Pressure Relief and Permeability Improvement

Variable diameter drilling integration equipment for drilling, caritation pressure relief and permeability improvement is an advanced equipment for coal seam mining. Through variable diameter drilling technology, coal seam cavitation, pressure relief and permeability improvement can be realized, and coal seam mining efficiency and safety can be improved. This unit will investigate the manufacturing company, structural composition, working principle and main application scenarios of the equipment, and explore its significance to China's energy exploitation.

At present, there are a number of manufacturing companies at home and abroad to produce variable diameter drilling, cavitation, relief pressure and permeability improvement integration equipment. Among them, the well-known domestic manufacturing companies include China Coal Technology Engineering Group, etc. Foreign manufacturing companies include Joy Global Inc. of the United States and Eickhoff Bergbautechnik GmbH of Germany, etc. These companies have high technical strength and market share in the field of coal mining equipment manufacturing.

Structure Composition

The integrated equipment of coal seam caving, pressure relief and permeability improvement in variable diameter drilling is mainly composed of drill rig, drill rod, drill bit and control system. The drill rig is the core component of the equipment, which is used to control the rotation and propulsion of the drill pipe to achieve the drilling operation. The drill rod is the part that connects the drill rig to the drill bit and is responsible for transmitting power and carrying loads during drilling. The drill bit is the working part of the equipment, which realizes the purpose of cavitation, pressure relief and permeability improvement by rotating and impacting the coal seam. The control system is used to monitor and control the operating status of the equipment to ensure the safety and stability of the operation.

KXJ-7300-500 mechanical coal seam cavitation pressure relief device is shown in Fig. 4-18.

Variable diameter mechanical cavitation device is shown in Fig. 4-19, Table 4-8.

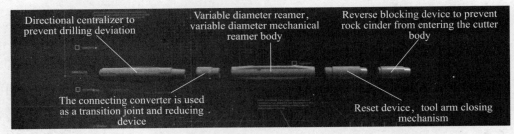

Fig. 4-18 KXJ-7300-500 mechanical coal seam cavitation pressure relief device

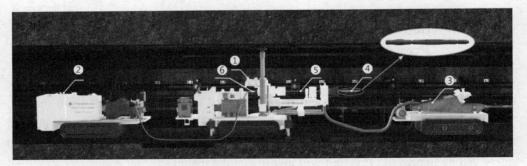

Fig. 4-19 Composition of variable diameter mechanical cavitation system

Table 4-8 Composition of a complete set of mechanical cavitation equipment

No.	Composition of a complete set of mechanical cavitation equipment
1	ZDY7300LX crawler full hydraulic tunnel drill for coal mining
2	BQWL315/16-XQ315/12 high pressure water pumping station
3	KFS-50/11 vibrating screen solid liquid separator for mining
4	Variable diameter mechanical cavitation device
5	Seal drill pipe with high pressure
6	Rotary joint with high pressure

Operation Principle

The working principle of the integrated equipment of variable diameter drilling for coal seam cavitation, pressure relief and permeabilty improvement is to realize the cavitation, pressure relief and permeability improvement of coal seam through drilling technology. The specific working process is as follows: first, the drill rod and drill bit are sent into the coal seam, and the drilling operation is carried out by rotation and propulsion; then, in the process of drilling, the drill bit rotates and impinges the coal seam to realize cavitation and pressure relief; finally, the equipment is monitored and controlled by the control system to ensure the safety and stability of the operation (Fig. 4-20).

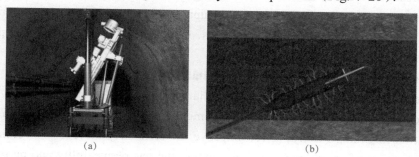

Fig. 4-20 Schematic diagram of mechanical coal seam cavitation construction

(a) The coal seam cavitation machine is in the tunnel; (b) Borehole diagram

Scope of Application

The mechanical coal seam cavitation device which can realize the integration of coal seam drilling and cavitation operation through the opening and closing of the mechanical knife arm. It can achieve pressure relief and permeability improvement of coal seam, improve the efficiency and concentration of gas extraction, and is widely used in gas extraction of perforating hole, coal seam and driving face.

Process Advantages

The outburst risk can be quickly eliminated by unloading and pumping. The driving rate of the cover lane through the layer drilling has been increased from the original 100 m/ month to 200 m/ month, and the pre-pumping of the bedding strip has been increased from the original less than 50 m/ month to 100 m/ month.

The equipment can expand the borehole aperture to 500-1, 200 mm, increase the permeability by 2-3 orders of magnitude, reduce the amount of gas extraction borehole engineering by 70%-80% under ideal conditions, increase the concentration by more than 70%, increase the amount of gas extraction by more than 7 times, shorten the extraction standard time by about 3 times, and reduce the cost of tons of coal gas treatment by 50%-60%.

Application Effect

(1) The gas concentration of cavitation holes is 1.5-3.4 times that of ordinary holes;

（2）The construction days of excavation single cycle are reduced from the original 10 days to 5 days;

（3）Gas extraction is 4.29-11.57 times that of ordinary drilling;

（4）The cavitation efficiency per meter is increased by 3-5 times.

Main Application Scenarios

The integrated equipment of variable diameter drilling for coal seam cavitation, pressure relief and permeability improvement is mainly used in coal seam cavitation, pressure relief and permeability improvement during coal seam mining. It can improve the efficiency and safety of coal seam mining and reduce the risk of accidents during coal seam mining. At the same time, the equipment can also be applied to other mining fields, such as metal mines, non-metal mines, etc., and has a wide range of application prospects.

Significance for China's Energy Exploitation

The integrated equipment of variable diameter drilling for coal seam cavitation, pressure relief and permeability improvement is of great significance to energy exploitation in China. Firstly, it can improve the efficiency of coal seam mining and reduce the waste of energy resources. Secondly, the gas permeability and permeability of coal seam can be improved through pressure relief and permeability improvement technology, and the recovery rate of coal seam can be increased. In addition, the equipment can also reduce the risk of accidents during coal seam mining and ensure the safety of miners. Therefore, it is of great economic and social benefit to popularize the integrated equipment of variable diameter drilling for coal seam cavitation, pressure relief and permeability improvement for China's energy exploitation.

【Words and phrases】

integration	n. 结合；融合；取消种族隔离；（数）积分法；求积分
cavitation	n. [流] 气穴现象；空穴作用；成穴
coal seam mining	煤层开采
coal seam caving	煤层放顶煤
efficiency	n. 效率；效能；（机器的）功率
investigate	v. 侦察（某事）；调查（某人）；研究
permeability	n. 渗透性；透磁率；磁导率；弥漫
rod	n. 棒；杆；钓竿
drill rod	钻杆
propulsion	n. 推进；推进力
impinge	v. 对……有明显作用（或影响）；妨碍；侵犯；侵占；撞击
gas extraction	气体提取

perforating	v. 在……上打孔；穿孔；刺穿
perforating hole	射孔
the bedding strip	层理带
borehole	n. 钻孔；井眼；（为探测石油或水）地上凿洞
aperture	n. 孔；穴；（照相机、望远镜等的）光圈；孔径；缝隙
the borehole aperture	钻孔孔径

Course practice

1. Complex sentence analysis.

（1）...the drilling operation is carried out by rotation and propulsion....

is carried out：被执行，指某个任务、计划或活动已经开始并正在进行。

【译文】……通过旋转和推进的方式进行钻孔操作……

（2）The construction days of excavation single cycle are reduced from the original 10 days to 5 days.

are reduced from ...：被减少，表示某物的数量、程度或力量已经降低。

【译文】掘进单周期施工天数由原来的 10 天减少为 5 天。

（3）...the recovery rate of coal seam can be increased.

...can be increased：……能被增加。

【译文】……可以提高煤层的复原率。

2. Translate the following sentences.

（1）The drill pipe is the working part that connects the drill to the drill bit and is responsible for transmitting power and carrying loads during drilling.

（2）The drill bit is the working part of the equipment, which realizes the purpose of cavitation, pressure relief and permeability improvement by rotating and impacting the coal seam.

3. Choose the proper answer to fill in the blank and translate the sentences.

（through, in, of, at, from, to）

（1）(　　) variable diameter drilling technology, coal seam caving, pressure relief and permeability improvement can be realized, and coal seam mining efficiency and safety can be improved.

（2）The integrated equipment of coal seam caving, pressure relief and permeability improvement (　　) variable diameter drilling is mainly composed (　　) drill rig, drill rod, drill bit and control system.

Translation of the text

隧道施工设备

可变径钻孔、造穴、卸压、增透一体化装备是一种用于煤层开采的先进设备，通过可变径钻孔技术，可实现对煤层的造穴、卸压和增透，提高煤层开采效率和安全性。本单元将对该装备的制造公司、结构组成、工作原理及主要应用场景进行调研，并探讨其对我国能源开采的意义。

目前，国内外有多家制造公司生产可变径钻孔、造穴、卸压、增透一体化装备。其中，国内知名制造公司包括中国煤炭科工集团等。国外制造公司包括美国久益环球公司、德国艾柯夫采矿技术有限公司等。这些公司在煤矿设备制造领域具有较强的技术实力和较高的市场份额。

结构组成

可变径钻孔、造穴、卸压、增透一体化装备主要由钻机、钻杆、钻头和控制系统组成。钻机是装备的核心部件，用于控制钻杆的旋转和推进，实现钻孔操作。钻杆是连接钻机和钻头的部件，负责传递动力和承受钻孔过程中的载荷。钻头是装备的工作部件，通过旋转和冲击煤层，实现造穴、卸压和增透的目的。控制系统用于监控和控制装备的运行状态，确保操作的安全性和稳定性。

KXJ-7300-500 机械式煤层造穴卸压装置如图 4-18 所示。

可变径机械造穴装置如图 4-19、表 4-8 所示。

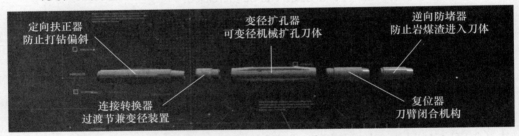

图 4-18　KXJ-7300-500 机械式煤层造穴卸压装置

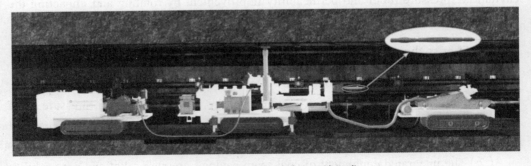

图 4-19　可变径机械造穴系统组成

表 4-8 机械式造穴成套设备组成

序号	机械式造穴成套设备组成
1	ZDY7300LX 煤矿用履带式全液压坑道钻机
2	BQWL315/16-XQ315/12 高压清水泵站
3	KFS-50/11 矿用振动筛式固液分离机
4	可变径机械造穴装置
5	高压密封钻杆
6	高压旋转接头

工作原理

可变径钻孔、造穴、卸压、增透一体化装备的工作原理是通过钻孔技术实现对煤层的造穴、卸压和增透。具体工作流程如下：首先，钻机将钻杆和钻头送入煤层，通过旋转和推进的方式进行钻孔操作；其次，在钻孔过程中钻头通过旋转和冲击煤层，实现对煤层的造穴和卸压；最后，通过控制系统对装备进行监控和控制，确保操作的安全性和稳定性（图 4-20）。

 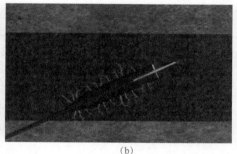

图 4-20 机械式煤层造穴施工示意

(a) 煤层造穴机在隧道中；(b) 钻孔示意

适用范围

机械式煤层造穴装置可以通过机械刀臂的开合实现煤层钻孔、造穴作业一体化的成套装置，可实现煤层卸压、增透，提高瓦斯抽采效率和浓度，广泛应用于射孔、煤层和掘进工作面的瓦斯抽采。

工艺优点

通过卸荷和泵送能快速消除瓦斯爆发的危险性。穿层钻孔掩护岩巷掘进由原来的 100 m/月提高到 200 m/月，层理带预抽由原来的不足 50 m/月提高到 100 m/月。

该装备可将钻孔孔径扩大到 500～1 200 mm，渗透率增大 2～3 个数量级，理想条件下瓦斯抽采钻孔工程量减少 70%～80%，浓度可提高 70% 以上，瓦斯抽采量可提高 7 倍以上，抽采达标时长缩短为原来的 1/3，吨煤瓦斯处理成本降低 50%～60%。

应用效果

（1）造穴孔的瓦斯浓度是普通钻孔的 1.5～3.4 倍；
（2）掘进单周期施工天数由原来的 10 天减少为 5 天；
（3）瓦斯抽采量是普通钻孔的 4.29～11.57 倍；
（4）每米造穴的效率提高了 3～5 倍。

主要应用场景

可变径钻孔、造穴、卸压、增透一体化装备主要应用于煤层开采过程中的煤层造穴、卸压和增透工作。它可以提高煤层开采的效率和安全性，减少煤层开采过程中的事故风险。同时，该装备还可以应用于其他矿山开采领域，如金属矿山、非金属矿山等，具有广泛的应用前景。

对我国能源开采的意义

可变径钻孔、造穴、卸压、增透一体化装备对我国能源开采具有重要意义。首先，它可以提高煤层开采的效率，减少能源、资源的浪费。其次，通过卸压和增透技术，可以改善煤层和瓦斯的渗透性，提高煤层的复原率。此外，该装备还可以减少煤层开采过程中的事故风险，保障矿工的安全。因此，推广应用可变径钻孔、造穴、卸压、增透一体化装备，对我国能源开采具有重要的经济和社会效益。

Evaluate

任务名称		隧道施工设备	姓名	组别	班级	学号	日期
考核内容及评分标准			分值	自评	组评	师评	平均分
三维目标	知识	了解可变径钻孔、造穴、卸压、增透一体化装备的组成及功能等相关知识	25 分				
	技能	能阅读专业英语文章、翻译专用英语词汇	40 分				
	素养	在学习过程中秉承科学精神、合作精神	35 分				
加分项	收获（10 分）	收获（借鉴、教训、改进等）：	你进步了吗？			加分	
			你帮助他人进步了吗？				
	问题（10 分）	发现问题、分析问题、解决方法、创新之处等：				加分	
总结与反思						总分	

3.2 Vehicle Radiator

 Text

Introduction of Typical Car Radiator

The vehicle radiator is an important heat dissipation device in vehicles such as cars and motorcycles, which is used to dissipate the heat generated by the engine and maintain the normal working temperature of the engine. With the continuous development of technology, new radiators such as assembled radiators, nitrogen shielded welding radiators and vacuum brazed radiators are gradually applied to the automobile and motorcycle manufacturing industry.

Vehicle Radiator Types and Manufacturing Companies

1. Assembled Radiator

An assembled radiator is a radiator that is manufactured separately from the radiator and the water tank, and finally assembled. Domestic manufacturing companies have are Shanghai Baosteel Group, Guangzhou Automobile Group, Beijing Automobile Group and so on. Foreign manufacturing companies have Germany Bosch, America Delphi, Japan Denso and so on.

2. Nitrogen Shielded Welding Radiator

Nitrogen shielded welding radiator is a kind Dmerica of radiator manufactured by nitrogen shielded welding technology. Domestic manufacturing companies are Shanghai Bao steel Group, Guangzhou Automobile Group, Beijing Automobile Group and so on. Foreign manufacturing companies have Germany Bosch, America Dmerica Delphi, Japan Denso and so on.

3. Vacuum Brazing Radiator

Vacuum brazing radiator is a kind of radiator made by vacuum brazing technology. Domestic manufacturing companies are Shanghai Bao steel Group, Guangzhou Automobile Group, Beijing Automobile Group and so on. Foreign manufacturing companies have Germany Bosch, America Delphi, Japan Denso and so on.

The Structure and Working Principle of the Radiator

1. Assembled Radiator

The assembled radiator consists of a heat sink and a water tank. The heat sink is usually made of aluminum alloy and has good thermal conductivity. The water tank is used to store the coolant. The working principle is that through the circulation of coolant, the heat generated by the engine is transferred to the heat sink, and then the amount is dissipated into the air through the blowing of the fan. The main application scenario is the engine cooling system of vehicles such as cars and motorcycles (Fig. 4-21, Fig. 4-22).

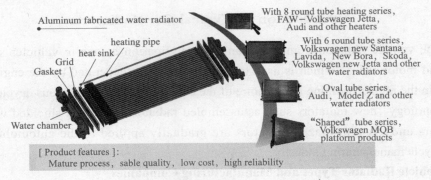

Fig. 4-21 Aluminum assembled water radiator

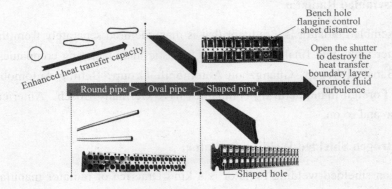

Fig. 4-22 Profile of water radiator

2. Nitrogen Shielded Welding Radiator

Nitrogen shielded welding radiator is made of nitrogen shielded welding technology, which has the advantages of high welding strength and good corrosion resistance. Its structure and working principle are similar to the assembly radiator, and the main application scenario is also the engine cooling system of vehicles such as automobiles and motorcycles.

3. Vacuum Brazing Radiator

Vacuum brazing radiator is made of vacuum brazing technology, which has the advantages of high welding strength and good sealing. Its structure and working principle are similar to the assembly radiator, and the main application scenario is also the engine cooling system of vehicles such as automobiles and motorcycles (Fig. 4-23).

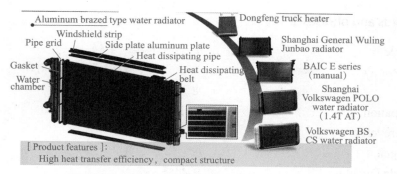

Fig. 4-23 Aluminum brazed type water radiator

The Significance of the Development of Radiator Industry to China's Automobile/Motorcycle Manufacturing Industry

As new types of radiators, prefabricated radiators, nitrogen shielded welding radiators and vacuum brazed radiators are of great significance to automobile/motorcycle manufacturing industry in China.

Firstly, these new radiators use advanced manufacturing technology with higher welding strength and tightness, which can improve the service life and reliability of the radiator, reduce the failure rate, and improve the overall performance of the car/motorcycle.

Secondly, the structure and working principle of these radiators are more advanced than traditional radiators, which can more effectively distribute the heat generated by the engine, improve the cooling effect of the engine, ensure the normal working temperature of the engine, and extend the service life of the engine.

Finally, the application of these new radiators has promoted the technological upgrading and industrial upgrading of China's automobile/motorcycle manufacturing industry, improved the competitiveness of China's automobile/motorcycle manufacturing industry, and promoted the development of the industry.

In summary, as a new type of radiator, assembled radiator, nitrogen shielded welding radiator and vacuum brazed radiator are produced by a number of manufacturing companies at home and abroad, and have important application scenarios and significance in the automobile/motorcycle manufacturing industry. The application of these radiators has promoted the technological progress and industrial upgrading of China's automobile/motorcycle manufacturing industry, improved the performance and reliability of the product, and promoted the development of the industry (Fig.4-24).

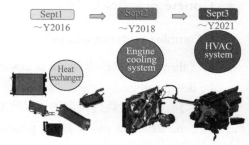

Fig. 4-24 Development stage of vehicle radiator

【Words and phrases】

vehicle	n. 交通工具；车辆
radiator	n. 暖气片；散热器；（暖气片型）燃油炉；电热器
dissipation	n. 浪费；消散
motorcycle	n. 摩托车
nitrogen	n. ［化学］氮
shielded welding	保护焊
nitrogen shielded welding radiator	氮保护焊散热器
vacuum	n. 真空；真空容器
brazed	adj. 钎焊的；铜焊的 v. 用铜焊接（braze 的过去式）
vacuum brazing radiator	真空钎焊散热器
water tank	水箱
domestic	adj. 家庭的；国内的；家用的
nitrogen protected welding radiator	氮保护焊接散热器
heat sink	散热器；热沉
aluminum	n. 铝
thermal conductivity	热导率
coolant	n. 冷却剂
circulation	n. 发行量；销售量；血液循环；流传；流通
corrosion	n. 腐蚀；侵蚀；腐蚀产生的物质
scenario	n. 设想，可能发生的情况；（电影、戏剧等的）剧情梗概；（艺术或文学作品中的）场景
the main application scenario	主要应用场景
sealing	n. 猎捕海豹；封闭；密封 v. 系紧；封牢
prefabricated radiator	预制散热器

Course practice

1. Complex sentence analysis.

（1）...then the amount of heat is dissipated into the air through the blowing of the fan.

...be dissipated into...：消散为……

【译文】……然后通过风扇的吹动将热量散发到空气中。

（2）...the structure and working principle of these radiators are more advanced than traditional radiators...

...more advanced than...：比……先进。

【译文】……这些散热器的结构和工作原理比传统散热器更先进……

2. Translate the following sentences.

(1) Vacuum brazing radiator is made of vacuum brazing technology, which has the advantages of high welding strength and good sealing.

(2) Its structure and working principle are similar to the assembly radiator, and the main application scenario is also the engine cooling system of vehicles such as automobiles and motorcycles.

3. Choose the proper answer to fill in the blank and translate the sentences.

(as, to, of, from, through)

(1) The main application scenario is also the engine cooling system of vehicles such (　　) automobiles and motorcycles.

(2) The water tank is used (　　) store the coolant.

Translation of the text

车载散热器

车载散热器是汽车和摩托车等交通工具中重要的散热设备,用于散发发动机产生的热量,保持发动机的正常工作温度。随着技术的不断发展,装配式散热器、氮气保护焊散热器和真空钎焊散热器等新型散热器逐渐应用于汽车和摩托车制造业。

车载散热器类型及其制造公司

1. 装配式散热器

装配式散热器是一种将散热片和水箱分开制造,最后进行装配的散热器。国内的制造公司有上海宝钢集团、广州汽车集团、北京汽车集团等。国外制造公司有德国博世、美国德尔福、日本电装等。

2. 氮气保护焊散热器

氮气保护焊散热器是一种采用氮气保护焊接技术的散热器。国内制造公司有上海宝钢集团、广州汽车集团、北京汽车集团等。国外制造公司有德国博世、美国德尔福、日本电装等。

3. 真空钎焊散热器

真空钎焊散热器是一种采用真空钎焊技术的散热器。国内制造公司有上海宝钢集团、广州汽车集团、北京汽车集团等。国外制造公司有德国博世、美国德尔福、日本电装等。

散热器的结构组成及工作原理

1. 装配式散热器

装配式散热器由散热片和水箱组成。散热片通常由铝合金制成,具有良好的导热性能，水箱用于储存冷却液。装配式散热器的工作原理是通过冷却液循环流动，将发动机产生的热量传递给散热片，然后通过风扇的吹动将热量散发到空气中。主要应用场景是汽车和摩托车等交通工具的发动机冷却系统（图4-21、图4-22）。

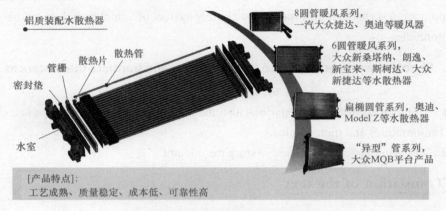

图 4-21 铝质装配式水散热器

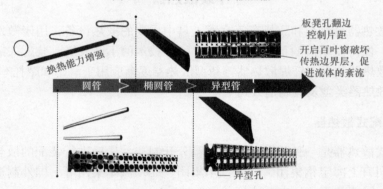

图 4-22 水散热器的型材

2. 氮气保护焊散热器

氮气保护焊散热器采用氮气保护焊接技术制造，具有焊接强度高、耐腐蚀性好等优点。其结构和工作原理与装配式散热器类似，主要应用场景也是汽车和摩托车等交通工具的发动机冷却系统。

3. 真空钎焊散热器

真空钎焊散热器采用真空钎焊技术制造，具有焊接强度高、密封性好等优点。其结构和工作原理与装配式散热器类似，主要应用场景也是汽车和摩托车等交通工具的发动机冷却系统（图4-23）。

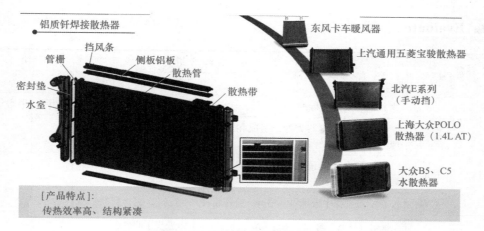

图 4-23 铝质钎焊接式水散热器

发展散热器制造对我国汽车、摩托车制造业的意义

装配式散热器、氮气保护焊散热器和真空钎焊散热器作为新型散热器，对我国汽车、摩托车制造业发展具有重要意义。

首先，这些新型散热器采用先进的制造技术，具有更高的焊接强度和密封性，能够提高散热器的使用寿命和可靠性，减少故障率，提高汽车、摩托车的整体性能。

其次，这些散热器的结构和工作原理比传统散热器更加先进，能够更有效地散发发动机产生的热量，提高发动机的冷却效果，保证发动机的正常工作温度，延长发动机的使用寿命。

最后，这些新型散热器的应用推动了我国汽车、摩托车制造业的技术升级和产业升级，提高了我国汽车/摩托车制造业的竞争力，促进了行业的发展。

综上所述，装配式散热器、氮气保护焊散热器和真空钎焊散热器作为新型散热器，在国内外都有多家制造公司生产，并且在汽车、摩托车制造业中具有重要的应用场景和意义。这些散热器的应用推动了我国汽车、摩托车制造业的技术进步和产业升级，提高了产品的性能和可靠性，促进了行业的发展（图 4-24）。

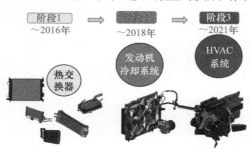

图 4-24 车载散热器的发展阶段

Evaluate

任务名称			车载散热器		姓名	组别	班级	学号	日期	
考核内容及评分标准					分值	自评	组评	师评	平均分	
三维目标		知识	了解车载散热器类型、组成和工作原理等相关知识		25分					
		技能	能阅读专业英语文章、翻译专用英语词汇		40分					
		素养	在学习过程中秉承科学精神、合作精神		35分					
加分项	收获（10分）		收获（借鉴、教训、改进等）：		你进步了吗？			加分		
					你帮助他人进步了吗？					
	问题（10分）		发现问题、分析问题、解决方法、创新之处等：					加分		
总结与反思								总分		

References

[1] 朱林，杨春杰. 机电工程专业英语［M］. 2版. 北京：北京大学出版社，2010.

[2] 石金艳，谢永超. 机电与数控专业英语（高职高专机械设计与制造专业规划教材）［M］. 北京：清华大学出版社，2014.

[3] 施平. 机械工程专业英语［M］. 15版. 哈尔滨：哈尔滨工业大学出版社，2014.

[4] 常红梅. 数控技术应用专业英语［M］. 2版. 北京：化学工业出版社，2012.

[5] 李娜，程尧. 机械工程专业英语［M］. 西安：西北工业大学出版社，2021.

[6] 林慧珠. 制冷与空调专业英语［M］. 3版. 北京：机械工业出版社，2015.

[7] 邓君香，孙暄. 飞机机电专业英语［M］. 北京：中国民航出版社，2015.

[8] Serra O. Advanced interpretation of wireline logs［M］. Paris：Schlumberger Limited，1986.

[9] Schlumberger Limited. Schlumberger Log interpretation volume I-principles［M］. Paris：Schlumberger Limited，1972.

[10] Edwardson M J, Girner H M, Parkison H R, et al. Calculation of formation temperature disturbances caused by mud circulation［J］. Jour. Pel. Tech, 1962, 14（4）：416-426.

[11] Alger R P. Interpretation of electric logs in fresh water wells in unconsolidated formations［M］. SPWLA Symposium, 1966.

[12] Archie G E. The electrical resistivity log as an aid in determining some reservoir characteristics［J］. Pet. Tech., 1942, 5（1）.

[13] Winsauer W O, Shearin H M, Masson P H, et al. Resistivity of brine saturated sands in relation to pore geometry［J］. AAPG Bulletin, 1952（36）：2.

[14] Archie G E. Classification of carbonate reservoir rocks and petrophysical considerations［J］. AAPG Bulletin, 1952（36）：2.

[15] Segesman F, Tixier M P. Some effects of invasion on the SP curve［J］. Jour. Pet. Tech., 1959.

[16] Threadgold P. Some problems and uncertainties in log lnterpretation［M］. SPWLA Symposium, 1971.

[17] Wyllie M R J, Gregory A R, Gardner G H F. An experimental investigation of factors affecting elastic wave velocities in porous media［J］. Geophysics, 1958, 23（3）. See also earlier article in Geophysics, 1956, 21（1）.

[18] Kunz K S, Moran J H. Some effects of formation anisotropy on resistivity measurements in boreholes [J]. Geophysics, 1958, 23 (4).

[19] 李秀民，孙大满，袁士宝. 石油工业俄语阅读教程 [M]. 东营：中国石油大学出版社，2008.

[20] 山东青州市南张石油机械厂. GJ 型供浆泵产品使用说明书 [Z]. 青州：南张石油机械厂，2009.

[21] 山东青州市南张石油机械厂. LW500-NY 型/LW400-NY 型钻井液离心分离机产品使用说明书 [Z]. 青州：南张石油机械厂，2009.

[22] 铁福来装备制造集团股份有限公司. 铁福来煤矿钻机智造专家 [Z]. 宝丰：铁福来装备制造集团股份有限公司，2021.

[23] 永红散热器公司（中航工业贵航股份）. 永红散热器公司简介 [Z]. 贵阳：永红散热器公司（中航工业贵航股份），2021.

[24] GD 光达传动. GD 光达传动 R/F/K/S 系列-齿轮减速电机宣传册 [Z]. 安顺：GD 光达传动，2021.